Ruth Cassungo Cruz

Implementation of a Virtual Environment as a support for the Mathematics ASAP

Ruth Cassungo Cruz

Implementation of a Virtual Environment as a support for the Mathematics ASAP

to 11th grade students of IPGB

ScienciaScripts

Imprint
Any brand names and product names mentioned in this book are subject to trademark, brand or patent protection and are trademarks or registered trademarks of their respective holders. The use of brand names, product names, common names, trade names, product descriptions etc. even without a particular marking in this work is in no way to be construed to mean that such names may be regarded as unrestricted in respect of trademark and brand protection legislation and could thus be used by anyone.

Cover image: www.ingimage.com

This book is a translation from the original published under ISBN 978-620-4-19588-9.

Publisher:
Sciencia Scripts
is a trademark of
Dodo Books Indian Ocean Ltd. and OmniScriptum S.R.L publishing group

120 High Road, East Finchley, London, N2 9ED, United Kingdom
Str. Armeneasca 28/1, office 1, Chisinau MD-2012, Republic of Moldova, Europe
Printed at: see last page
ISBN: 978-620-5-58914-4

Copyright © Ruth Cassungo Cruz
Copyright © 2023 Dodo Books Indian Ocean Ltd. and OmniScriptum S.R.L publishing group

Contents

DEDICATOR 2
THANKS..3
COMMITMENT BY THE AUTHOR 4
SUMMARY 5
CHAPTER 1 6
CHAPTER 2 13
CHAPTER 3 34
CHAPTER 4 53
CHAPTER 5 73
Conclusions..93
Recommendations 101
BIBLIOGRAPHY..102
ANNEXES 119

DEDICATOR

I dedicate this merit to God and to the memory of my father

"Joao Maria Cassungo" may his soul rest in peace.

THANKS

I thank God for giving me the wisdom and perseverance to achieve this purpose.

To **Roberto Cruz Acosta,** my husband, for the great support and for$a$on he gave me in the elaboration of this work.

To my tutor, **Prof.ª Dr.ª Viviane Sartori,** for her availability in guiding the preparation of the dissertation.

To my parents for the for$a to achieve my dreams, for the affection and support.

To my children **Luchano Antonio and Ramona Cruz,** for understanding my constant absences during the research period.

To my brothers for their encouragement and affection.

I would like to thank all my teachers who, through their lectures and knowledge, have helped in my academic and professional training.

I would also like to thank all those people that directly or indirectly have always been helpful, influencing me with their experiences. I must not forget the FUNIBER's direction, for the information provided in the administrative and academic field during the formation period.

COMMITMENT BY THE AUTHOR

I, Ruth Bulica Chana Cassungo Cruz, declare that:

The content of this document is a reflection of my personal work and I declare that before any notification of plagiarism, copy or lack of the original source, I am directly responsible for the legal, economical and administrative situation, without affecting the supervisor of the work, the University and all the institutes that collaborated in the referred work, assuming the consequences derived from such practices.

Signature:

SUMMARY

The present dissertation aims to contribute to the development of mathematical competences, with the implementation of a virtual learning environment, through the moodle platform: used to create *online* teaching spaces and to manage, distribute and control all the non presential formative activities with the intention of improving the learning processes, contributing as a support to the Mathematics discipline and facilitating the general and specific development of the students of the eleventh grade of the IPGB. It is based on the low results that have been presented by the students, when presenting internal and external tests related to this reasoning and on the need to improve these performances. The sample size was 60 students chosen at random, to whom an evaluation instrument of the study variables was applied. To develop this work, the mixed approach was used, since both approaches are mixed in most of their stages, and therefore, it is convenient to combine them to obtain information that allows triangulation. The aim of this study is the implementation of a virtual learning environment to support the teaching and learning process of Mathematics. The research carried out took into account the diagnosis made by the Ministry of Education of Benguela province, taking into account, in addition, all the specialized literature review regarding virtual learning environments to support the subject of mathematics; A review of all the legal norms and regulations and the laws that approve the incorporation of technologies in education was also carried out; finally, a field research was carried out to know the learning process of mathematics and the teachers' management in the IPGB. The research allowed that the contribution for the integration was evaluated as positive in the PEA through the virtual environment of the IPGB. It resulted in the design and construction of a didactic proposal according to the population's conduction and to the national mathematical references, with adequate support for the development of random thinking in students of the eleventh grade. A personalized learning experience is highlighted, improving the mathematical competencies of the students from a learning model based on interaction, promoted by the mediation of computer resources, motivating a high level of satisfaction of teachers and students and approaching the contents, both in the representation of concepts and the development of skills in problem solving for a significant learning. In the IPGB, there are possibilities of continuing research for other researchers who will be able to deepen the pedagogical studies, regarding the benefits of the ICT tools for the development of effective and efficient competencies, since the studies developed so far are few, and these experiences have not been disseminated to promote debate and socialization.

Keywords

Virtual Learning Environment, Didactic Design, Mathematics Learning, ICT, Pedagogy and Didactics.

CHAPTER 1
INTRODUCTION.

The current society presents new challenges for education, one of them is the impact of technologies, mainly in subjects like mathematics, the way they are influencing the students' learning, diversifying different ways of understanding and assimilation of contents, where the teacher plays the role of a facilitator in this process.

The complexity of this subject in its teaching and learning is a difficulty for the educational system and a challenge for teachers and students when dealing with abstract entities and algorithms where apprehension is systematically required, being the use of abstract thinking one of the most difficult subjects of the curriculum at the level of General Basic Education (EGB), Baccalaureate and Higher Education Mendoza and Mosquera (2006).

For Gamboa (2007) the virtual learning environments (VLE) allow the teacher to be the one who guides the student in the understanding of learning through activities that will allow him/her to integrate with the other participants, having a more active participation in his/her own learning with different study and learning modalities.

With the insertion of ICTs, the teacher has new challenges, because, although it is true that there are different ways of taking knowledge to students, the teacher must master this new environment in order to be the content facilitator in the construction of students' knowledge, allowing them to develop skills, where the student goes from passive learner to active in their educational and social performance.

Being the main concern of the teachers the assimilation and understanding of the contents of the mathematics subject, the great variety of tools that ICT provides, gives the necessary support so that the teacher can improve the techniques and learning methods, giving the possibility of carrying out the activities in the classroom.

Technologies in education and especially in the subject of mathematics motivate students to carry out searches, discover new concepts unknown to them until then and improve the skills of interaction and communication between students and teacher Arias Cabezas et al. (2008), Cruz and Puentes (2012).

To develop meaningful learning and achieve the desired skills, two fundamental pillars are needed:

* ICTs are indeed a resource, but the use made of them by the teacher is fundamental for the results to be achieved.

* ICTs are the teacher's means of achieving their objectives. Therefore, it is essential not to make the most common mistake, teaching students to use ICT when in reality it would be how to achieve mathematical content using ICT as a tool.

In the development of today's society, the teaching of mathematics is fundamental, which is why educational systems pay special attention to its learning, although society itself is changing, traditional teaching methods still prevail, which are emphasised in procedures that little stimulate the opportunity to reflect on the processes that are carried out.

Gamboa (2007), Martin Cantoral (2001) and Martin (2000) believe that using technologies in mathematics teaching can break the usual routine that students are used to.

This work has the following theme: Virtual Learning Environment as a support for the Mathematics discipline, where students will have access to the information from anywhere, allowing them to enhance their learning. The title of this work is the following: **Implementation of a virtual environment as support to the teaching and learning process of Mathematics for students of the eleventh grade of the Instituto Politecnico da Grapa, in Benguela.**

1.1 Empirical justification of the research.

The world today is living in a new era, that of knowledge, where ICT has played a fundamental role in the development of society. Technologies have transformed a globalised world, facilitating the economy, communications; education cannot remain oblivious to these changes in which society has been transformed and must provide answers to the problems arising in it. Using technologies to improve new learning and the construction of knowledge.

How can we approach the evolutionary process of technologies in education without incorporating them? For today's world is not possible, the school must go hand in hand with technological development, reaching it through the improvement of programs and project proposals that bring the school closer to current changes. Therefore, it is not enough to arrive, it is necessary to have a teaching staff with updated didactic and pedagogical projects, both structural and pedagogical, that stimulate the use of technologies and the development of competencies in significant learning.

The subject of mathematics is recognised as being difficult by the students and complex by the teachers; at the Polytechnic Institute of Gra?a in Benguela, traditionalist learning still prevails, and often the barriers are mental, falsely believing that help will come in the form of some miracle. Without realizing that change must begin with the teachers themselves, there must be a concern with how to facilitate student learning, and not be content to accept that the discipline is difficult or that it would be complex to carry out projects that stimulate the way of teaching and learning, both for teachers and students.

The results of the last three years were not the best and yet no initiative was presented to transform this panorama. Technologies can be a solution, provided that the teacher is prepared and uses them to achieve understanding and assimilation of the content.

The diagnoses carried out by the Benguela Provincial Directorate of Education, Science and Technology, during the year 2018, obtained better results than in previous years, even so, 50% of the students in the eleventh grade of secondary education in the province obtained results below 50% of the desired score, indicating that the medium technicians still show little mastery of mathematical skills.

The Instituto Politecnico da Gra?a in Benguela, in particular, obtained the lowest score in the province, with a difference of 84.5% in relation to the other students with the highest score.

These results show the differences between secondary school students in Benguela province and the consequences of this reality for university studies.

During the year 2018 at the Instituto Politecnico da Graga in Benguela (IPGB), the average obtained at the end of the semester is equivalent to 31/50; and only 50% in the Mathematics subject. These data are a reflection of the path taken in mathematics by the students of IPGB, demonstrating the accumulated deficiencies in the development of abstract thinking skills, being essential skills for the development of professional skills.

With this situation, teachers who teach mathematics in grade 10 face a problem in the teaching and learning of mathematics in students who have low attention span, concentration, as well as weak abstract reasoning processes, not allowing them to assimilate mathematical contents efficiently.

Facing this problem of learning mathematics for students of the eleventh grade of the Instituto Politecnico da Graga in Benguela (IPGB), the teachers could not count on the instruments and processes that facilitate the mediation of the learning of concepts, theorems and algorithmic processes, being contents of this study area, which improve the mathematical abilities of the students.

From an observation made by the author of this research and mathematics teacher of the 11th grade of the Instituto Politecnico da Graga in Benguela (IPGB), to know how has been the use of the technologies in the educational process, it was detected that several teachers present difficulties in the computer applications in the use of the classrooms in a didactic way, making the classrooms with traditional methods, putting in risk the quality of the classes. Therefore, it was verified that there are still difficulties in the use of technologies in the teaching-learning process, as shown below:

1. Insufficient use of technologies in the teaching-learning process.
2. Insufficient use of tools that generate learning resources.
3. Little use of the school intranet in the development of the educational teaching process.
4. Insufficient methodological planning in the use of technologies in the elaboration of computer resources that mediate the educational teaching process.

This research will develop a Virtual Learning Environment with the objective of supporting the subject of Mathematics in students of the IPGB 11th grade, using technologies as mediators in learning learning, allowing students to build their own learning. The teacher responsible for the virtual environment will be the facilitator of the same, performing activities, publishing information of interest, clarifying doubts about the subjects, assigning tasks.

This research proposes to apply a virtual learning environment in support of the subject of mathematics for a better understanding by the students of this subject, taking into account the various tools that VLEs offer, and that can help the teacher to monitor the learning development of each student, allowing the teacher to administer, upload content, publish information of interest related to the subject, plan tasks, discussion forums on different topics and perform an evaluation of them, and also allows answering tasks to maintain permanent communication with all students.

1.2 Formulation of the problem:

How to implement the use of Virtual Learning Environments as a support tool for the Mathematics subject of the eleventh grade students of the Instituto Politecnico da Gra?a in Benguela (IPGB), through virtual courses?

1.3 Research questions:

1. What are the theoretical-methodological references that should sustain the didactic design of a VLE, as a support tool for the Mathematics subject?

2. What are the characteristics of the didactic design of a VLE, as a support tool for the Mathematics subject for students of the eleventh grade of the Instituto Politecnico da Gra?a in Benguela (IPGB)?

3. What components and relates should the didactic design of a VLE have as a support tool for the Mathematics subject of the eleventh grade students of the Instituto Politecnico da Gra?a in Benguela (IPGB)?

4. How to evaluate the contribution of the VLE didactic design, as a support tool to the Mathematics discipline of students of the eleventh grade of the Instituto Politecnico da Gra?a in Benguela (IPGB)?

1.4. General objective.

Elaborate a didactic design of a Virtual Learning Environment, as a support tool for the subject of Mathematics for students of the eleventh grade of the Instituto Politecnico da Gra?a in Benguela.

1.4.1. *Specific objectives:*

1. To analyse the main theoretical-methodological references that support the didactic design of a VLE, as a support tool for the subject of Mathematics.

2. Review the bibliographic material of the Ministry of National Education, (curriculum guidelines, competency norms), that allows to support the development of the project.

3. To characterize the didactic design of the VLE as a support tool for the Mathematics subject of the students of the eleventh grade of the Instituto Politecnico da Gra?a in Benguela (IPGB).

4. To evaluate and appropriate different resources and technological tools that improve the teaching-learning processes and develop students' random mathematical thinking.

5. To develop the didactic design of a VLE as a support tool for the Mathematics subject of the eleventh grade students of the Instituto Politecnico da Gra?a in Benguela (IPGB).

6. To verify the practical contribution of the didactic design of a VLE as a support tool for the Mathematics subject of the students of the eleventh grade of the Instituto Politecnico da Gra?a in Benguela (IPGB).

1.5. Main theoretical aspects assumed to address the problem.

Within the areas of knowledge, mathematics is one of the disciplines used in several of them; for many students it is difficult to understand and assimilate its contents, being one of the disciplines with lower qualifications when compared to other disciplines. Marchesi, Coll and Palacios (1990) consider that at certain times students feel frustrated in learning mathematics, which causes them to condition their professional selection to avoid being overwhelmed by their problems.

The teaching of mathematics has great challenges, on the one hand there is the complexity of the subject which is recognised by teachers and on the other hand students recognise that the content is difficult to understand and assimilate, many due to predispositions and other times due to a deficit in the learning of the subject in previous classes, which has still not been overcome. How to achieve the construction of the students' own knowledge? What would be the abilities and methodologies to contribute to a significant learning that allows the construction of knowledge in students? For Barros (2008): it is important that the student learns in a perspective of logical sense so that he/she can relate his/her cognitive structure, generating a dynamic different from the usual and allowing new ideas in the students.

It is recognised by teachers and students that face-to-face classrooms often lack the nuance of innovation that would motivate students to develop the capacity to wonder or observe, they rarely get to experiment, they lack creativity, making classrooms tedious; problems are long and difficult, far from collaboration; which is attributed to a large number of student rejections Hidalgo, Marato and Palacios (2004). This situation has been solved in different ways, currently the use of technologies has helped to a better understanding of the subject, allowing the students to use different technological tools that have facilitated a better understanding and comprehension in problem solving, allowing them to develop skills as well as the construction of their own knowledge.

These difficulties generate a rejection on the part of the students to mathematics, adding many times in which they avoid activities directed to the area of study, this panorama of learning difficulties is complemented by the conditions of the education in Angola, where the spatial reasoning is a subject little discussed by students of secondary education. Often in the traditional perspective where it is difficult for students to visualize space more coherent with geometric elements, leaving it as something abstract and meaningless.

These difficulties demand a new approach in the teaching-learning process of mathematics, where there is little or no interactivity with the technologies, little training on the didactics of informatics in the mathematics discipline, the teachers do not consider the ICT as useful as mediators and therefore reject its use in the teaching-learning process.

1.6. Social relevance.

To elaborate a didactic design of a Virtual Learning Environment as a mediator in the Mathematics discipline for students of the 11th grade of the Instituto Politecnico da Graga in

Benguela. The creation and diversification of didactic materials that will be used in the cognitive processes, in the interaction between other students, achieving a greater dynamics of openness in the construction of their own knowledge, the individual and collective work will generate in the students a greater autonomy allowing them to manage their learning in a more planned way, developing abilities for the management of knowledge. This will also lead to a more dynamic, creative and innovative management of the teacher, with the creation of new educational models that involve the current needs of society, creating a new working environment between the student and the teacher, facilitating the interaction between the students and the teacher as a facilitator in student learning.

1.7. Practical implications.

The scientific novelty of the investigation resides in the didactic design of a VLE to integrate the subject of Mathematics and support the students of the eleventh grade of the Instituto Politecnico da Graga in Benguela in the mediation of the teaching and learning of Mathematics. It will allow a better understanding of the mathematical content through the Virtual Learning Environment.

1.8. Methodological usefulness.

In this investigation, the methods used were the following:

- **Historical-logical**: this method was used to know the antecedents related to this investigative theme, as well as the most current tendencies related to the object of study and the field of action of this investigation.

- **Systemic approach**: to design the proposed VLE as a system that determines its perspectives, components and relationships.

- **Analysis - synthesis and induction - deduction**: to determine generalities and specificities in the object of study and in the field of action; as well as in the theoretical basis and elaboration of the didactic design of the VLE.

Empirical methods were used in the form of interviews with students, teachers and administrators, and observation in order to get to know the educational teaching process in the classroom and to see the development of both the teacher and the students in their environment. The document analysis helped to find out all the documentation relating to the study programmes of the discipline and the internal regulations applied in the institution and at provincial and national level, and to know the current situation at international level. Furthermore, focus group techniques were applied, in which it was possible to collect first hand information from students and teachers using the technique of Iadov. The statisticians and mathematicians helped with the descriptive statistics, and a tool such as Excel 2009 was used for the processing of all the data obtained in these applied methods and empirical techniques.

The theoretical contribution is achieved with the didactic design of the VLE integrating the mathematics discipline. However, the practical contribution is reflected in:

- The implementation of the VLE teaching project, which is aimed at teachers and specialists in the IPGB as mediators in the process of teaching and learning mathematics.
- The implementation of a VLE for the subject of Mathematics, developed on the Moodle platform, according to the didactic design of the proposed VLE and following the implementation guide.

The dissertation is structured in five chapters:

Chapter I - provides an overview of the work, the motivation, the contribution, the research problem, the theme, the scientific questions and the unit of analysis of the research work, the need and justification of the problem, the objectives, the tasks, the purpose and the limitations of the research are presented. A brief description of the other chapters that make up the memoir is also made.

Chapter II - Develops the theoretical and methodological foundations for the didactic design of a VLE for the subject of Mathematics. Furthermore, it describes the Virtual Learning Environment and the Mathematics subject;

Chapter III - Details the methodology used in the investigation, population and sample, as well as the data, techniques and instruments applied in this investigation.

Chapter IV - Responds to the general objective of the investigation; approaches the analysis of needs, the definition of the proposed objectives, the selection of contents and the evaluation that was used in the proposal: Didactic Design of a Virtual Learning Environment, as a support tool for the Mathematics subject of the students of the eleventh grade of the Polytechnic Institute of Graga in Benguela.

Chapter V - The results and interpretation of the evaluation and contribution of the proposal are presented: Didactic Design of a Virtual Learning Environment, as a support tool for the subject of Mathematics for students of the eleventh grade of the Instituto Politecnico da Graga in Benguela, responding to the proposed objectives.

At the end, the final conclusions are also dedicated, based on the results of Chapter V, adding perspectives and some recommendations for future research.

At the end of the paper, the bibliography and appendices are added.

CHAPTER 2
VIRTUAL ENVIRONMENT FOR LEARNING THE SUBJECT OF MATHEMATICS.

This chapter develops the theoretical and methodological foundations that should support the didactic design of the VLE for the subject of mathematics. Furthermore, the teaching-learning media and the role of ICT and its transformation are analysed. Afterwards, the evolution of the current tendencies of VLEs is investigated. It continues with the analysis of the tendencies and challenges of the teaching-learning process in the discipline of Mathematics; as well as the use of ICT in the didactic link of mathematical-methodological strategies that involve ICT. It concludes with the analysis of two types of reasoning that arise in students when solving mathematical problems and interacting in a virtual environment.

2.1. Information and Communication Technologies (ICT) and the means of teaching and learning.

In the knowledge era, it is necessary to recognize how technologies have been incorporated to human development, they have become the main tool as a way to access knowledge, the way of communication, the transmission of information has been consolidated as never before. From the educational perspective, it becomes necessary to use technologies as mediators in the teaching-learning processes. It is not new in the educational field to use mediating tools long before ICT, even though many of them are still in use; nowadays, technologies play a transforming role due to all the possibilities they offer.

For Vygotsky (1987), there is a relationship between subject and object where the dialectical interaction that occurs between them is mediated by socio-cultural instruments. For a better understanding, the influence of the context in which the individual moves and the elements used by each one are recognised.

The concept of mediation is related to the tools used, according to Vygotsky (1987), the tools are linked to the same human development which man uses to achieve different goals and transform his environment with them. Vygotsky recognizes that the working tools used by man transform nature in time and space. Therefore, their use fulfills the function of mediator.

The use of the tools in different contexts is used as mediation, which happens in a primary way in which the subject, through the tool, is represented by him. A second context in which the subject uses this mediation by means of signs which Vygotsky considers as "psychological tools" versus "physical tools", always oriented in a relationship between man and his context, Vygotsky (1987). All this activity is carried out from a mental plane in which the subject captures, interprets, internalizes and reports experiences previously acquired from his perspective.

That is why these mediations between man and society are the instruments between people and their culture. The signs of writing, as well as art and language, have provided man with a context and a socio-cultural experience in order to assimilate different forms of human activities and to be able to understand both the material and spiritual means developed by this culture.

Man, in his need to communicate, has used these signs as part of his social behavior, which have allowed him to have a psychological influence on internal activities that have developed quantitatively in psychological processes, taking them to a higher stage. These psychological processes have come to form a system of social relations, where the communication established between the subject and others has allowed a joint activity, and to develop their psychological functions, forming part of the individual activity process of each subject.

Therefore, each subject has at his disposal tools and signs of communication within the sociocultural context where an interaction is applied in all the social and cultural processes, as well as in the psychological processes of each one of them. This places the tool as a mediating process in the completion of the psychological processes carried out both inside and outside the school and the teaching and learning processes.

From the Vygotskyan point of view, the process of teaching and learning options beyond their speeches and all the means used for teaching in this context, the schools use their means of communication for the dissemination of the students and their families. The technologies which constitute a protagonist within the media in this communication are those which can bring about changes in the behaviour of man Lima (2005, p. 6).

In the teaching-learning process, with the integration of ICT, two essential relates are distinguished Frias (2007, p. 27):

• The dialogues between the subject and the digital didactic material mediated by a communicative interface. The asynchronous dialogue between the student and his relation with the materials (subject-object relation, mediated by other subjects).

• The different channels of communication allow a more social, direct exchange with the community. A dialogue is a synchronous and asynchronous interaction influencing positions, expectations and behaviour.

When communication technologies are used as support in the teaching-learning process, a technological mediation is generated, where the tool will be the one to mediate between the parties and facilitate the acquisition of new knowledge. Therefore, its main objective is to facilitate the interactions, improving communication, besides promoting several ways of interaction. Many of these dialogues are carried out in a pedagogical way, since they facilitate the construction of knowledge. The teacher as facilitator uses technology as a mediator between the subject that builds knowledge and the object of knowledge. This means that the teacher creates activities, tasks, forums, chats, generates contents with which the students will achieve a greater interpersonal dialogue teacher-student or student-student.

The pedagogical mediation happens individually and collectively, when learning is promoted in teaching institutions, and it provides the student when he collaborates, given that he discovers and rediscovers himself; and when he facilitates intercommunication. According to Lima (2005, p.7), pedagogical mediation is a procedure through which the teacher guides his actions taking into account the dialogue; in other words, to achieve the objectives defined in a preliminary way, the collaboration of the students is necessary, and only in this way, will allow them to reveal certain precise competences for social life.

These technologies have given a revolutionary boost to society, as Fandos (2009a and 2009b) explains, creating the students' characteristic
"Schneider (2002) cited by Guiza (2011) creating 3 evolutionary processes in the use of ICT in education:

- This learning using the computer appeared between 1980 and 1995, where in a first phase what was sought was to explore the software and create simulators in which parameters were operated within a dynamic system using hypertexts.
- For 1996 and 2000, the popularity of computer networks or better known as www (World Wide Web) for the facilities in communication such as e-mail, interaction, information search and computer based training, receive more attention.
- From 2001 to the present day, communication is mediated by computers, mobile phones, tablets, achieving a subject-subject interaction where they receive more attention. The web consolidates itself as the dominant technology in collaborative work and in the exchange of experiences and knowledge.

Castellanos et al. (2009) and Vygotsky (2001) agree that the collaborative is a characteristic of this type of learning, where the collaborative has been the essence of a component for participation and internationalisation. And in the case of Anaya (2004), Collazo (2004) and Frias (2008), they agree that the technologies can break barriers and help in the temporal space in education.

Authors such as Anaya (2004), Cemile (2008), Er (2009), Guiza (2011), Izquierdo and Pardo (2005), James - Gordon (2007), Lakkala (2010), Lee (2011), Mondejar et al. (2007), Perez (2002), Sanchez et al. (2008), Silva (2007), Vazquez (2011) and Villasevil (2009), there has been a debate regarding the influence of technologies in their evolution as a means of teaching and learning within education, considering them as auxiliary means of teaching, despite the fact that they have been gaining an increasingly larger space within the teaching and learning process.

In order to better understand this topic, it is necessary to know how technologies have been mediated within teaching and learning. The different material and technical characteristics of the different components of the teaching and learning process are highlighted. Not forgetting the didactic characteristics, as well as the objectives and the contents and methods to be used, taking into account how these components will be complemented.

For Alvarez (1996, p. 59) there are different mechanisms of operation in this process where the methods are manifested and expressed attending the material objects within teaching and learning where it is related to the rest of the components: the problem, the objective, the object, the content, the organization and the finalization with the result. According to Alvarez (1999), this concession is made in the forms in which the thought of signs and symbols are exteriorized through language.

In the case of Addine (2004), he mentions how these means of teaching and learning are related to other components of the ASP. By way of example: the students and the teaching group. Enabling the media to reach the manifestation of thought in the SAP participants, influencing the way they communicate and collaborate in their learning.

For Area (1998) and Fandos (2009a and 2009b) recognise learning media as "technological resources". Addine and Garda (2009), Collazo (2009), Fernandez and Parra (2004), Gonzalez et al. (2004a, 2004b) and Silvestre and Zilberstein (2003) confirm this criterion in which they consider technologies as mediators in their performance and skills.

For Gonzalez (1986, p. 67) technologies fulfil certain functions within teaching and learning, such as communication and integration between the different components. According to Collazo (2009), these means of communication acquire a relevance in the transmission and appropriation of information, experimenting a treatment in the control of the teaching and learning processes.

These positions coincide in the sense that the medium is a component of the SAE, in close relation with the rest, carrier of contents, material support of the method and mediator of teaching and learning.

Area (1998), Cabero (1996), Del Touro (2006) and Fernandez (2005), these authors believe that for the design, the use and evaluation of these teaching and learning media are made from three dimensions: semantics, syntactics and practice. The first referred to what the medium says; the second, to how it is presented; and the last, to how and for what it is used.

For, Anaya (2004), Cemile (2008), Er (2009), Guiza (2011), Izquierdo and Pardo (2005), James - Gordon (2007), Lakkala (2010), Lee (2011), Mondejar et al. (2007), Perez (2002), Sanchez et al. (2008), Silva (2007), Vazquez (2011) and Villasevil (2009), in which the great diversity that facilitates the technologies in education, generated the need to restructure the system in which the evaluation is approached, taking into account the semantic, syntactic and practical dimensions.

Guiza (2011, p. 54) understands ICTs as the "set of technologies that enable the acquisition, production, storage, processing, recording and presentation of information in sound, image and data formats...[which] include electronics as their basic technology, which supports the development of telecommunications, information technology and audio-visual".

Fandos (2003a, 2003b) technologies have made hypermedia and multimedia possible, diversifying the possibilities of information. This has enabled services such as e-mail, chat, discussion forums, podcasts, vodcasts and others, which have transformed the way in which communication occurs between people globally.

Frias (2008), Lakkala (2010) and Villasevil (2009), coincide in the creation of technological resources that benefit education and break with the traditional means of teaching and learning, achieving a better understanding, communication, reaching higher levels. Rodriguez (2008) and Silva (2007) are of the opinion that with these technological resources, communication activities and understanding of content are much better than traditional means of teaching and learning.

Faced with these challenges, we need competent teachers capable of using this pedagogical mediation in which the presence of these technologies is composed of hardware, software, audiovisuals, but above all, to be able to process all of this didactically in an integral way with the teaching and learning process. Because this implies teachers with a technological culture capable of transforming the educational reality. This is possible by applying knowledge in procedural techniques, intuition and creative imagination to didactically process the ICT resources according to the objectives.

Pedagogical mediation with technology opens new horizons in the search for procedures on how to use information, appropriating aesthetic and playful possibilities present in any creation. The integration of the teaching and learning process is the inspiration for teachers.

In this new perspective, the activity of the teacher does not disappear, on the contrary, it becomes more active, in the elaboration and later didactic processing, becoming a content facilitator. In this way, it guides the learning with activities, materials, in which the students appropriate those that come to them mediated through the technologies and the interaction with ICT resources, and other users.

2.2. Virtual learning environments: evolution and current trends and their didactic design.

The virtual learning environments are recognized as a development software that allows not only the assignment of contents, but also of different materials combined so that the student can use them for a better understanding of the same. Virtual learning environments are a tool used by teachers in the management of contents and materials to improve the learning comprehension of certain contents, facilitating the communication between students and students, besides improving the communication between teacher and students, allowing this communication to take place outside the traditional context of an online course. Once inside this learning environment, the teacher can control the whole learning process and the performance of each student, allowing the teacher to offer a differentiated attention taking into account the problems that may arise, which achieves to guarantee the effectiveness of the teaching and learning process.

For Santos and Okada (2003), the VLEs can be defined as web environments used by teachers and students, communicators, for the development and synchronous and asynchronous integration geographically distant. The VLEs offer several tools that allow performing different activities facilitating access to content and proposals of certain subjects, as well as other resources.

With the internet, it is possible to access the VLE courses, many are the ones that use it from companies, universities, schools, but it is possible that the VLEs are seen as products specifically for education. To assist the teacher in the administration of course content and the management of content for students, allowing the monitoring of student development in the process of teaching and learning.

According to Almeida (2003), it means:

1. They allow you to propose activities that facilitate meaningful learning, as well as being able to plan them.
2. Create multiple materials, media and languages.
3. The student has a facilitator.
4. Allows you to search for information.
5. It helps to reflect on the processes and;
6. It helps to formalise concepts.

These proposals are truly revolutionary in the educational field, being positive and of great interest for the teaching and learning processes. These didactic tendencies have managed to merge, adapt and improve thanks to technologies, creating a universe still to be discovered and creating new alternatives, which have managed to break down barriers in the use of virtuality applied to education.

Therefore, Duart and Sangra (2010, p. 7), explain that there are doubts about the use and the possibilities of didactics in education, considering that the methodologies that are normally used in traditional processes within the classroom must be improved to be applied in a totally virtual context. These learning processes receive a totally different treatment within the context of virtual tools, considering that they go through a transformation process in which these virtual tools have to comply with the educational purposes, the objectives, the sequences of the methodologies that are designed to comply and implement the assessments within the virtual environment, supporting the teaching and learning process, thus transforming the role that the teacher plays within this new environment. In this way it is possible for the student to pass from a passive state to an active state inside his own learning with the use of the tools inside the virtual learning environment, the student can not only interact with the content but also it will allow him to construct his own knowledge through the tools with which he will work and it will facilitate his self evaluation of the self learning, generating a critical thought not only of the possible mistakes made but also to analyse the resolution of problems that other students make and in which they can also issue their own criteria, facilitating an interaction about the contents of the mathematics subject. For Lara (2001, p. 133), there are different theories in the teaching of learning in virtual environments. In this respect, the main controversy is related between conductivism and constructivism within the virtual learning environment in which many authors have issued their criteria and they suggest the use of both theories, taking advantage of the mixed strategies that each one has.

As, Lara (2001, p. 134) explains:

The behavioural perspective should be used mainly to deal with organisational aspects, such as defining the structure of the process, setting objectives and managing evaluations. The constructivist perspective should be used to manage highly academic aspects, such as the definition of interaction strategies and the definition of individual and group activities that contribute to the achievement of the objectives.

Among the various didactic tendencies of virtual learning environments, the theory of conversation, the theory of knowledge and the connectivist theory stand out. This theory breaks with classical didactics and with the teaching and learning processes, being fundamental the responsibility that both the student has in the construction of learning and the teacher as a mediator and facilitator of contents through technologies.

Apparently, these innovations of virtual education had implicit: objectives, goals and purposes of a form of education immersed in the virtual context are openly unknown. These styles of practice measured by technologies and the teaching and learning process remain tacit, where each teacher acts in his or her own particular way, bringing with it a discussion of these trends.

So that many are inquisitively asking themselves about the different positions and didactic tendencies of virtual education, how useful they can really contribute to the EAP. These new discussions about the possibilities of virtual tendencies generate a search for the necessity and importance of: what results are obtained when interpreting different didactic tendencies in virtual education?

Several authors express it in different ways, as María Lourdes Rincon, Venezuelan, shows a conceptualization and importance of Virtual Learning Environments, as a very useful tool in academic advising, its use is a key factor in the changes produced in these visual environments. She also offers several alternatives and scenarios on how to face and learn in this information society.

For Marianela Magro Fernandez-Arlyne Solano Gonzalez, from Costa Rica, didactic strategies can be proposed for virtual learning environments if they are applied in courses that are implemented to these strategies are categorised into 3 types: centred on individual learning; group learning - emphasising the collaboration between them, taking into account the tools within the virtual learning environments platform.

In the case of Cocunubo-Suarez et al. (2018), Colombian, even an assessment in virtual environments taking into account its usefulness should be recognised based on the ISO 9126, 14598 and 25000-SquaRE standards, where they reviewed a series of certain documents with useful topics for the assessment, from which the characteristics integrated in the 25000-SQuaRE were identified with greater frequency.

Keyler Rodriguez Velazquez, Juan Miguel Perez Fauria, Geisi Torre Garda, carried out significant research in virtual learning environments where 20 demonstrate in these investigations the use of these technologies in teaching and learning processes; being of great impact in didactics for a better understanding of the contents by the students. Within this type of research highlights the great importance of the requirements in the pedagogical practice using technologies. Adell and Castaneda (2012, p. 15) consider emerging pedagogies, known as pedagogical trends. According to Castro (2004) and Nunez (1999), the development of science has been increasingly inclined towards the integration of knowledge, taking into account the facilities that it offers both to the teacher, in the administration, design and control of student activities, and for the students to participate and construct their own knowledge, providing them with new abilities that they did not have before and offering them an interaction in a context where they can interact with other students. Pickett and Cadenasso (2002) consider that this concept is transferable among other sciences exposing components among which the community is found, referring to the interaction among them, the communication, the individuality of each one of the participants, considering the self-learning by the facilities offered by this environment, and the processes in which students and teachers develop and participate, bringing with them a common benefit.

For Uden and Domiani (2007) in this learning context a learning community is created and its organisation and management play a fundamental role in the results to be obtained. For these ecosystems represent a solution for the complex problems which are present nowadays in the industrial academic sector. Authors such as Nikolaidou et al. (2009) consider that the providers in this content communication depend on the technology infrastructure.

Alvarez et al. (2011), Alvarez et al. (2012), Alvarez and Rodriguez (2011), Nikolaidou et al. (2009) and Uden and Domiani (2007), these authors agree that learning ecosystems are characterized by a common and organized infrastructure; allowing interconnectivity between other ecosystems in which they mutually benefit from participation.

Alvarez et al. (2012, p. 11) consider ecosystems in a transdisciplinary way, using strategies that benefit organizations in order to articulate virtual and face-to-face spaces with the participation of students, teachers, collaborators, involving the entire academic community and seeking beyond its borders people of common interest.Therefore, they consider that this "localization of the ecosystem clashes ... with the reductionist and disciplinary visions of the curriculum", to the detriment of educational practices with ICT.

For Unigarro (2004), he recognizes the VLEs where an educational process is generated, an educational communicative action is produced, interaction of the content and construction of knowledge of the formative intentions being different from a class, allowing an asynchronous and synchronous temporality, through the teacher-student and student-student relationship, called cyberspace.

Herrera (2006a and 2006b) the VLEs are able to provide the conditions for the realization of learning activities, characterizing them as conceptual and constitutive elements. They refer to the interaction established in the virtual design, texts, images, and everything related to what can stimulate the senses and the psychological aspects, are found in the social environment and in the collaborative work.

Taking into account what was proposed above, Virtual Learning Environment is understood as a system with technological scenarios, constituting an educational context structured and determined by didactic foundations and principles, offering the possibility to manage and evolve both technically and pedagogically, by means of strategies and didactically, allowing those who participate to communicate and collaborate synchronously and asynchronously.

Del Toro (2006, p. 42) many of the authors propose the need for a virtual learning environment where the teaching and learning process will depend mainly on the hyperenvironments from its semantic and syntactic dimensions and from the practice considering its interrelations. Rodriguez (2008, p. 34) is in agreement with Del Toro (2006), who considers that the didactic designs of virtual environments must be constituted by a coherent structure with the formative units and activities in execution to a specific purpose.

Collazo (2004), the processes of didactic design of virtual courses, understands it as a process of production, considering it as insufficiencies in the field of ICT, which still persist in teachers, because they do not have the overcoming of teachers, moreover, so little is used for these purposes.

If we consider previous definitions such as those of Adams and De Vaney (2009), Athanasios (2007), Cartelli et al. (2008), Cemile (2008), Chard (2011), Gomez (2002), Lee (2011) and Skelton (2007), we would recognize that didactic design in virtual learning environments has different aspects to consider:

▫ It must be based on "theoretical and methodological foundations", since it must comply with categories, norms, principles and structure and relates, allowing the anticipated representation of the VLEs that will be obtained.

▫ These processes are necessary in the construction of virtual learning environments, considering the actors who will participate in it. That is why this research aims at a didactic design of a virtual environment where the theoretical foundations and methodological requirements in the design of the virtual learning environment take into account the criteria of the teachers who teach the subject of mathematics, as well as students and other people who are committed to this project, taking into account that they serve as a basis in this productive process and its overcoming.

The didactic design of a VLE presents the requirements that should guide its construction Paquette (2002) cited by Guiza (2011, p. 93) proposes the following:

1. Availability at any time and place the student needs.
2. Allows the student to manage their learning.

For Ladyshewsky (2004) quoted by Nisanci (2005, p. 25) proposes the following requirements as follows:
1. Possibility of immediate feedback from subject - subject and within the group.
2. Providing the skills that the student needs to understand the contents of the mathematics subject and giving the teacher greater control of each student's activities.

Silva (2007, p. 139), in turn, establishes those that follow:
1. Represent the social space explicitly.
2. Facilitate mechanisms for participants to be active, and also to build the virtual space.
3. It allows both face-to-face and distance learning to be enriching for the PEA.
4. The integration of technology in the administration, management, distribution, monitoring and evaluation of collaborative work.

Lakkala (2010, p. 82) similarly alludes to the following requirements:
1. It enables the integration of face-to-face activities mediated by technology.
2. Provides support on research challenges.
3. It offers participants opportunities for communication, interactivity, interaction with collaborative work.

Callaghan et al. (2009), Gao et al. (2009), Pavon (2008) and Thorsteinsson and Denton (2008), agree that this is not always fulfilled in educational practice. It is acknowledged that most of the above-mentioned requirements comply with the technology-related printpios where addressed in the literature, creating a deficit in others that deepen the psycho-pedagogical elements necessary for didactic design. For this reason, the requirements below are added:
- To offer technological facilities taking into consideration the individual cognitive-affective characteristics of the participants.
- To recognise the pedagogical model of the institution(s) and the modality of the EAP for which the VLE is implemented.
- These mechanisms serve as feedback, allowing participants to develop in guiding their learning.
- All participants should receive psycho-pedagogical support in addition to technological support, which should gradually decrease until mastery is achieved.
-In the virtual learning environments, both the pedagogical and the technological aspects should be taken into consideration, facilitating the conception of contents following international standards.
- To specify a didactic strategy that enables the development of the EAP through the VLE in its entirety (design, implementation and evaluation).

Perez (2002), considers that the didactic design of a virtual learning environment is limited to recreational and non-educational purposes, designed with a single content. According to Sanchez et al. (2008) the didactic designs of virtual learning environments are structured with resources and activities that facilitate communication, evaluation and use of tools by students and better control of teachers. Er (2009) considers that the management of learning should include a syllabus with notes, action forums and assignments, as well as the necessary tools to be able to use them to solve problems.

For Villasevil (2009) the didactic design has the student-teacher platform and the database server. The platform is formed by the basic design: generalized information for the student (student data, notice board and announcements), information supported by a theme (forums, chats and web links) and the most frequent questions and answers.

Fandos (2003a and 2003b) proposes a structure with videoconferences and multimedia where notes can be taken as a starting point to distribute both doubts and queries, so that it can serve as a means of information, such as e-mail, distribution list, as well as forums - group work chat, creating a virtual platform where different activities can be carried out and can be controlled by the teacher.

According to Er (2009), Fandos (2003a and 2003b), Perez (2002), Sanchez et al. (2008) and Villasevil (2009), these authors consider that the structure of technologies to create a virtual learning environment should answer specific questions: what for, why use technologies? How to use them? How to transform its management and its evolution? How to create its structure for a better understanding? Among many others.

It is recognised that the didactic design of the VLE should contain:

1. "Technological tools" for the presentation and management of content, communication, as well as for evaluation and monitoring.

2. "Didactic materials" their contents within the disciplines of mathematics in question.

Another group of researchers has another proposal of didactic design with the pedagogical aspect. James - Gordon (2007) a design based on six categories in the form of components, namely: content and learning requirements (and objectives), philosophy, delivery, management, technology and VLE.

With the use of technologies in the management of a didactic design of the virtual learning environment, several authors were consulted. One of them Vazquez (2011) considers a technological proposal considering the tools of this technology and the contribution of licenses. On the other hand, Fantini (2011) believes that virtual learning environments should be constituted by an administration subsystem.

(IMS, 2003) considers 3 levels of learning technology systems (Learning Technology System Architecture), where the user is the student, the assessment and control by the teacher, and the facility to assign content to the platform by creating a data storage management system.

In this sense, the project reinforces the tools that should be used in the VLE. Considering the possible actors (teacher and student in this case) as personal components that the environment will have. But the existence of limitations is rooted in the types of relationships that the participants will have, considering the teacher only as a software system.

These pedagogical models that focus the virtual learning environments are decisive for the didactic design of them, where several authors such as Anaya (2004) recognizes that there is in their fundamental systems to comment the educational model of technological teaching, besides considering that the technological platforms in the management of the teaching-learning processes are positive. Silva (2007) agrees with this, as he states that these elements of didactic design in virtual learning environments represent a pedagogical model in which teachers use them to improve the teaching and learning process.

It has great importance, because these authors point out the dependence of the VLE on the pedagogical model. However, Anaya (2004) finds differences between the media and the Internet, a position that is not shared in this thesis, since he exposes that the Internet is the technologies that can be considered as a medium, which offers ambiguity and lack of consistency in its didactic design. Silva (2007) does not explain the internal structure and the relations that determine the pedagogical model.

Several authors consider that these scenarios where the didactic designs applied to these logical platforms represent a technology mediated education. Alfonso et al. (2006) consider that the pedagogical models, the didactic materials within these didactic strategies represent the development of the teaching and learning processes. Rodriguez (2008) considers that the components must be interrelated in the didactic strategies of development of the teaching and learning process between teachers-students and the didactic material.

Lakkala (2010) proposes that the pedagogical infrastructure within the interrelated framework should consider the technical, the social, the cognitive and the epistemological. Thus, combining the social and the cognitive in a didactic strategy that represents the virtual learning environment in this pedagogical model and the participants of the didactic materials as an epistemological component.

The didactic designs proposed by authors Alfonso et al. (2006), Lakkala (2010) and Rodriguez (2008) consider other components for the didactic design of virtual learning environments:

1. "Space or stage" allows a VLE organisation to communicate collaborative work synchronously and asynchronously.

2. "Didactic strategy" with specific actions and steps to follow in the design, implementation and evaluation of the EAP in the VLE.

Del Toro (2006) proposes that "in the educational context" and "as educational contexts" are themselves. This is an extremely important element for the didactic design of the VLE. Taking into account these two proposals, external links are produced. The first determines the "VLE as an educational context". And the other produces a reaction in the opposite direction, with feedback or transformation, in the use of the VLE in the EAP, which will produce modifications in the "VLE in the educational context".

According to James - Gordon (2007, p. 39) this author considers that they really offer solutions that integrate these learning factors. Besides, he adds that the development of these technological and pedagogical aspects ignores the pedagogical role of learning; as far as it limits the methods that are used in virtual learning environments with certain technologies in a spedific moment and where it can be appreciated little integration among them.

For Collazo (2004, p. 85) defines that the "integral process that comprises the design, realization - assembly, revision - correction, validation, legalization of the course and teachers' supervision". The criterion of this author is shared with regard to the teachers' overcoming in didactically design a VLE.

James-Gordon (2007) says that the didactic design should serve as a basis for implementation in the structure and relationships of the didactic design. Alvarez et al. (2012) and Uden and Damiani (2007) carry out a diagnosis to identify the needs, from this the design and its implementation is elaborated, to later evaluate and finally value the impact of the VLE.

For some actors, the production of a virtual learning environment must be strongly linked to the institution fulfilling the performances applied of each teaching unit, its objectives and its activities, in addition to the evaluation, Guiza (2011). For Rodriguez (2008), the priority is in the design of the virtual learning environment considering the different dimensions as the theory of methodology and practice, concluding with an educational practice that is evaluated.

Therefore, it is concluded that the proposal of the virtual environment in its production process should exceed the expectations created, Collazo (2004), since it meets the stages designed for assembly and a pilot test carried out where teachers were evaluated throughout this process.

This does not mean that virtual learning environments are a magic solution that will solve all problems, on the contrary, it will depend mainly on the teacher, his role as facilitator of this learning, using technological tools will be fundamental for the success of this mission. Callaghan et al. (2009) argues that it is necessary to achieve an integration of virtual environments worldwide. Alvarez et al. (2012, p. 11) consider that virtual learning environments can become something really useful for learning and for the acquisition of the necessary skills both in communication and in the understanding of new knowledge, enriching both linguistic and cultural barriers. Considering James-Gordon (2007), we still recognize the existence of some limitations in the activities and works developed in some companies that offer virtual learning environments as a communication and collaboration proposal between them and their members.

2.3. Virtual learning environment: a tool to support mathematics teaching.

The means continue to be a component of the PEA, which allows us to reflect that its usefulness is not exclusive, but depends on the relates made by the other components, if we start from the strategies that the teacher carries out, these will act as the basis of the relates and will be the possible reach of the PEA, since the educational means must be part of the systematic, interactive and dynamic communicative process where knowledge and experiences flow, consolidating the interactions and the interpersonal relations Fernandez (2005) and Rico et al. (2002).

For Salinas (2002), if the above is taken into account, an analysis of the availability of the means must be carried out, taking into account the economic and technological feasibility, as well as a didactic feasibility study. According to the reality where it is carried out, the components of the SAP should be articulated, allowing a timely response to the demands of today's society.

Horrutinier (2006) is even clearer on this subject, the priority is to raise the mastery of the means of communication in teaching, fundamentally the technological ones, achieving this integration in this perspective consists of the condition of a system in which the elements that compose it act reciprocally. Not from an isolated event, nor from a teaching context with the simple condition of adding new teaching means to this process, managing to expose multiple challenges and questions about why, why and how to use the teaching means in an integrated way in this process.

Castaneda and Fernandez Dan Alaiza (2002), Garcia-Valcarcel and Gonzalez (2011), Horrutinier (2006), Lombillo (2011) and Lopez (2007), for these authors it is essential that the teacher can detect the need for the application of technologies so that students can opt for this type of tools that help them to face the difficulties and barriers that they encounter in a traditional classroom, since the understanding of the content of mathematics itself is recognized as having a high degree of difficulty. This is why, in order to achieve a transformation in the current needs that exist in the mathematics discipline, the need to implement virtual learning environments is recognised so that students can face the challenges during this process Horrutinier (2006, p. 116 and 118).

Considering the above, it should be understood that technologies in educational institutions can play an important role in the development of the teaching and learning process, given that they can be used as a means to build a memory of content of the didactic system between the subjects that are used, as well as to take into account the needs that are met and are known by teachers in professional practice, Lopez (2005, p. 5). Therefore, it is necessary to recognize that the learning of mathematics needs a relationship inside the classroom and outside it with the students, because it does not require only the learning of fundamental concepts or necessary procedures, in fact, difficulties are generated in which teachers will play a more active role in order to obtain successful results, in this sense, the didactics of mathematics studies its teaching process with the aim of understanding its problems and being able to solve them in order to strengthen the learning of students Serrano (1993).

When we want to disseminate the mathematical knowledge, the didactics of mathematics should be taken into consideration, because it defines parameters and communication processes for teaching, due to the fact that mathematical activities are different from other activities D'Amore (2006). In relation to this approach, it should be understood that the methodology enables the design of strategies, as well as the use of tools that help the student in the appropriation of conceptual and symbolic structures of mathematics.

For D'Amore (2006), the science of mathematics contemplates different contents of study during the processes of transmission and acquisition of the same, considering that mathematics is not always taught in the same way, since each teacher can add part of his or her personality in this performance, but without neglecting that the student must be motivated in the learning of mathematics. That is why, throughout the evolutionary process of the human being, the use of different communication and information tools is recognised, Jaramillo (2007). In the current reality, we can already count on digital resources and applications that are getting better every day, which are applied by the whole society, both in work environments and participating in the educational culture itself in the use of these tools. Knowing these technological tools in the field of education that support educational processes through virtual learning environments, Jaramillo (2007).

For this reason, in different works of academic investigation, the integration of technologies with the SAPs of mathematics are of transcendental importance. When it comes to changing the methodologies of the SAP, many researches on the teaching of mathematics are aimed at supporting this learning process through interactive environments and multimedia resources, Gallegos and Pena (2012). When these resources are used in this way, they are aimed at training in skills and attitudes that can have a positive impact on the social sphere.

The great diversity of computer programs that exist today and that can be useful in the accomplishment of educational works, as for example: HotPotatoes, Derive, Geogebra, considering the variety and quantity of mathematical contents that can be used in the form of hypertexts, images, graphs, applets, etc., many times known both by students and teachers, but despite the recognition of their usefulness, they still continue without incorporating them in their training process, creating resistance to change.

Technologies present a wide range of tools that can help and provide good opportunities for students to build and further develop their mathematical knowledge. An example of these are the dynamic softwares that favour dynamic representations of mathematical objects. Some heuristics such as the measurement of attributes (lengths, areas, perimeters), the use of CAS systems (Computer Algebra Systems) such as calculators help with algebraic operations and with aspects related to the meaning or interpretation of the results. John (1980), expresses some currents within behaviourism, but all share the following four elements:

a) The responses of the organism will depend on the situation, taking into account the object of study and the behaviour.

b) This being the case it will be subjective, if the method is absolutely empirical.

c) There are three pillars: where the situation is considered, but responses are expected and the organism is taken into account.

d) The purpose of psychology as an applied science is to modify behaviour.

"Behaviorism Science has conducted studies on human behavior, and is the philosophy of this science" Skinner (1990, p. 136).

In the 1980s, Lopez (2007, p. 65) states:

The application of these technologies in the teaching of mathematics during this decade was widely developed, through programs for performing and solving equations, strategy games showing a better way of relating teachers and students. The technology has become an important element for mathematics. With the strategy games, the students have improved their reasoning and have acquired reflective skills and have begun to feel motivated to learn and understand the subject of mathematics.

In the last two decades, attempts have already been made to improve education, and the teaching of mathematics in particular has predominated as a strong trend the incorporation of technologies in the school environment. Being a support tool providing a better learning Sunkel (2006, p. 57).

The study of mathematics has gone through many transformations throughout history, where it has suffered many changes in the teaching of mathematics before the emergence of technologies, considering that the learning of mathematics has been influenced mainly by the needs that the human being has taken into account in the understanding of his environment and the economic needs in which he is inserted and subject, considering that the knowledge is being built day by day through appropriate practices. For Ausebel (1918, p.63), this is not something that could be transmitted, in reality it is quite the opposite, knowledge must be constructed and reconstructed every day, the subject always facing each problem the results may change and may find different answers and this is only achieved if it is understood that one learns with each action that is performed.

Confirming the above, it could be said that the daily experiences provide knowledge not only when one participates in an action, but also when one listens, one also learns from the world that surrounds the human being, that is, education in its process of human formation meets the demands of the social being, taking into account the paradigms, categories and epistemological bases within a specific scientific framework. In view of what Arnal (1992, p. 98) argues, the concept of a critical theoretical idea represents a social science which is not purely empirical, it has only been interactive, which contributes by preventing in the studies the participative investigation.

The above is understood if we consider that knowledge is constructed from individual and not group interests, reaching it through the formation of subjects of participation and social transformation. What for the socio-critical paradigm is based on social ethics, accompanied with a reflexive character in a constructivist way. This self-reflection is carried out so that each subject becomes conscious within the group of what his/her role is; that is why the proposal is in the ideological cynic, in order to better understand this situation, psychoanalytical procedures are carried out for each individual, discovering his/her interests by means of the cynic. Achieving the construction and reconstruction of knowledge with a process between theory and practice.

An analysis of the technologies in the teaching of mathematics would help to understand the situation.

For Infante Quinteiro and Logreira (2010, p. 39):

Technology should be a factor or transversal axis of mathematics education. Consequently, there is a need to reaffirm curricula, pedagogical methods, and the relationship with society on the part of mathematics education from the new

information and communication technologies.

In order to understand this better, it is necessary to recognise the strong link between mathematics and technology, in which mathematics is recognised as the basis of all the sciences; Recognising the usefulness that the technologies represent in the world of mathematics, this means the need for changes both in the curricular aspect of mathematics education, and in the resources necessary to be able to carry out classrooms with the required quality, in which well-equipped laboratories are expected, with computer programmes and that are linked to the subject of mathematics, with Internet capacity, using techniques and methods in the process of teaching and learning mathematics.

When we think about technologies, the first thing that comes to mind is the use of computers, internet, etc. But we forget that such an old tool is part of the teaching-learning process and that it is still used today, the calculator. The use of technological resources in mathematics should be accepted with an open mind, in a practical, intelligent, but radical way for Angolan society. If we want to have men of sciences, it is necessary to strengthen the study of mathematics, which should be more linked to the technologies to improve the teaching-learning process of mathematics.

In this process, the teacher plays an important role, because he/she will be the one to face these transformations, guiding each stage and making them a reality, applying the technologies of mathematics, he/she will be the one responsible for achieving this interaction and for guaranteeing that the student has positive results, because he/she will be a facilitator, but also a negotiator and will have to make decisions, being the only person who knows when the technology should be applied. Taking into account that he is the one who best knows his students, what their weaknesses, strengths or needs are, therefore he knows when to apply a decision since he will know when he will obtain the best results.

With the technologies, the role of the teacher does not disappear, on the contrary, their performance is improved, achieving a better control of their students, helping and facilitating their learning, giving the possibility that each student is able to participate in the construction of their own learning. Besides preparing all the didactic content so that the students have a better performance showing their potential. Since technologies do not substitute the work of the teacher, they serve as a support tool, facilitating the work of the teacher and in turn facilitating a better assimilation and understanding of the content, guaranteeing that the student has the possibility to participate in the construction of knowledge that he or she will take care of, and the other students will also take care of it.

The technology forms a student with knowledge, critical and creative, able to reason, verify and solve problems that arise in their daily life. Preparing them for professional life as society is becoming more and more technologically integrated.

These technological devices continue to be of great help in the process of learning mathematics, for example:

- Calculators;
- Calculation software;
- Dynamic geometry programs.

All these resources require a computer for their execution, the majority of the students already know and dominate the technologies, and it is the social needs that more demands them with the technological abilities, and it is not the school that is in the first line, and depriving of the initiative in this context.

The use of technologies in mathematics helps to improve the academic performance of students in mathematics classes, not only by supporting, but also by creating, motivating and developing basic skills, representation, argumentation, thinking and reasoning in mathematics.

2.4 Mathematical skills developed in virtual learning environments.

Teachers and students recognize the importance of mathematics in their lives, with which many problems can be solved. Nevertheless, its complexity is known, so it is necessary to work with pedagogical strategies for a better assimilation and understanding in practice Benitez (2011) and Vasquez (2002).

Therefore, it is necessary to develop competences that allow solving problems in each particular context, taking into account the psychosocial resources where skills and attitudes are assigned:

a) the capacity for analysis and smthesis;
b) the ability to learn;
c) problem-solving skills;
d) the ability to apply knowledge;
e) the ability to drive digital technologies;
f) the skills to manage information and
g) the ability to work autonomously and in groups.

These competences can be combined with all the areas of knowledge, especially the study of mathematics. To understand what is the performance of mathematics in the world and the individual's ability to identify the competences of mathematics when the needs of life present themselves as a constructive, committed and reflective citizen ISEI-IVEI (2004).

With the use of VLEs, the diversification of didactic material used in the SAP increases, developing new options of interaction between people, managing to create more openness in the educational dynamics for the subject of mathematics.

With the fundamentals of improving mathematical performance and mathematical skills, there is the analogy of exposing communication, logical reasoning and problem solving to improve mathematical performance. Of course, none of this can be achieved by itself, it requires VLEs with problematic, meaningful and detailed situations that can achieve increasingly complex skills. Therefore, VLEs need didactic units with their methodological strategies in order to improve the EAP in the area of mathematics, allowing the development of both general and specific skills.

Therefore, it is necessary that these proposals generate the need for students to be able to perform the logical reasoning, through the solution of problems related to the construction of Virtual Learning Objects (VLE), create application proposals for mathematical products. To achieve these mathematical objects, the context should be taken into account, providing the student with the possibility of performing its interpretation, interacting with it, for a better understanding and application.

When mathematical competences are taken into account, it is a sign that the students' skills in logical reasoning, analysis, communication used in the process of solving mathematical problems in a variety of situations are being recognised INECSE (2005).

This problem-solving process is taken into consideration at different stages:

a) reach and identify the variables that are in the problem;
b) represent the problems from different points of view;
c) recognise the relationships that exist between the variables of the problem;
d) establish relationships between the representations made;
e) identify mathematics that can solve problems in everyday life;
f) relate the problem to a simpler one;
g) use a mathematical model to represent the problem;
h) justify the results and;
i) communicate the process and the solution.

According to Parnafes and Disessa (2004), the logical reasoning of each student has a different representation of each problem, which stands out in aspects of certain concepts, in which there are different representations, and in which a flexible understanding of each of them is achieved. This is why it is recognised that when relationships are established between students, cognitive processes are produced during problem solving. These situations require communication and interaction in this training process, generating different points of view in this cognitive process in which knowledge is derived from different points of reasoning when solving problems.

For his part, Ainsworth (2006) recognises the benefits of using a representation, getting at how information is encoded in the representation and understanding its relationship to the domain it represents. In this case, take into account that students want to select a representation that is meaningful to them.

The cognitive process itself represents a complex cognitive process and this helps to solve the problem. What additional skills are related? With the use of ICTs, does the individual have to develop? In this sense, Garcia and Benitez (2009) mention that with the incorporation of technologies, teachers must reaffirm the practice, fostering the guidance of students, taking into account the design of appropriate activities.

Nowadays, the pace of life is accelerated, so it can be determined that the contribution of virtual learning environments in the development of mathematical skills assists in better reasoning of problem solving, causing through the activities and tasks proposed by the teacher an academic performance in mathematics, which provides the student with a more active perception in the use of tools and in the understanding of this mathematical discipline. Cordoba (2015) and Gordillo (2017) state, "This is in contrast to techniques in which the student has to face strategies that are characterized by a memoristic approach.

Therefore, it represents a responsibility of both public and private institutions in the country that must generate strategies to guide the planning processes of both development planning and formative assessment in the area of mathematics, taking into account the learning needs and the skills that students must acquire in the meaningful learning of mathematics, this means that in the design of the same should consider the use of various tools in the didactic development of this subject MEN (2016) . For these authors, Rozo and Perez (2014) consider that virtual learning environments, as well as the different types of existing software, can improve learning and understanding of the content of mathematics, strengthening both the skills and the concepts related to this area of knowledge.

Hence the need to create a didactic strategy in the construction of a virtual learning environment to support the learning of mathematics. It should be considered what kind of tools would be appropriate in the learning of mathematics? Which would be the tools to be applied in these strategies that could strengthen the mathematical logical reasoning? How can we elaborate different activities so that students can develop logical reasoning in the resolution of problems that arise in the subject? In compliance with this, a bibliographic review on this theme was carried out, allowing the identification and analysis of the contribution of virtual learning environments in the competences of mathematics and logical reasoning in problem solving.

These didactic strategies are conceived taking into account both the skills and methods as well as the techniques and resources which have been planned in order to help the student to achieve significant learning and to facilitate the work of the teacher in controlling the performance of each student.

For Benedict (2000, p. 112), teaching strategies are conceived as part of a plan to achieve the objectives proposed during the teaching and learning process.

Before the teacher in the design of his plans for the teaching of mathematics creates strategies that lead the student to discover and rediscover contents that can explore different ways in the resolution of problems, achieving an interaction between the new knowledge and the already learned, where they will be able to face daily problems, within these pedagogical practices, taking into account that it must be useful not only in the planned classroom, but also as part of the didactic unit and the study programmes that are presented in this mathematical discipline.

The creation of different teaching strategies recognising the interactions that exist in the educational processes, in the content selected in the curricular units and in the materials to be used, independently of the different psychological and social theories that may exist in the learning of mathematics. When planning is decided upon, teachers are responsible for leading that process and making decisions about how, what and when to teach.

Therefore, teachers and other educational designers should reflect on the educational purposes, the choice, organisation and logical systematisation of contents, in addition to devising teaching strategies to be implemented in circumstances in which their pedagogical practices fit.

Nowadays, the teaching of mathematics "instrumental aspect" is, without a doubt, the science that helps with its knowledge to understand other areas of science, besides enriching other areas of knowledge in an interdisciplinary approach, always searching for alternatives to these new situations. It is necessary that teachers, managers and other people involved assume the importance of their actions in the success of the students.

CHAPTER 3
METHODOLOGY

3.1. Methodological approach.

Implementation of a virtual learning environment to support the learning of mathematics in the 11th grade of the Instituto Politecnico da Graga in Benguela (IPGB), located in the province of Benguela. This is an exploratory descriptive and correlational investigation, with a mixed methodology.

3.2. Type of study.

Different types of surveys were applied to meet the objectives proposed for each of them:

Exploratory because it is an unknown theme, little diffused and new in the context of the research. **Descriptive**: the analysis of the characteristics and usability in the learning of mathematics during the use of the virtual learning environment platform was carried out, as well as the teacher's management, its optimization in the face of these new tasks and its application with the technologies to be used in the near future. **Correlational**: the study was carried out taking into account the connections and the links between the different phenomena that occurred during this investigation. **Explanatory**: the causes of the phenomena and situations of the problem were sought as its main objective, being the reason for this investigation.

We analyzed quantitative data on the variables and the qualitative data complemented the quantitative data, which means that we worked with mixed research, that is, quantitative-qualitative. Many researchers, when using qualitative research, carry out narrative accounts of the phenomena studied and the techniques applied, such as the separation of participants and structured interviews. The main difference between the two methodologies, both quantitative to study the relationship and association between quantified variables, and qualitative, lies between the structural and situational contexts. Qualitative research recognises the depth of the context in which it is investigated, therefore it is able to identify which realities occur during this research, establishing a system of relationships and a dynamic structure. For quantitative research, it is capable of determining the associated or related forces between the variables, being able to generalise the results besides being objective in the inferences of a population from which all the samples come from. When studying the association and correlation, the intention is to make an inference that can explain that everything happens or not in a certain way.

Qualitative research is able to describe the experiences of daily life and give subjective meaning; it is able to recognise both the events, norms and values that are produced from the perspective of the object of study, therefore, the experience of the subject must be taken into account. Therefore, it is of great interest when it is necessary to know and understand the human experiences when feeling certain realities such as pain, healing, motivation, power or comfort.

In qualitative research one can find some of the meanings that the researcher has generated through the results of the interviewees, for example, the perceptions and motivations of each subject under investigation being at the centre of the analytical conclusions.

All research in which the subject participates as part of the process is part of a qualitative research in which it must be considered and applied as an integral part of the projects in which this type of research is developed, bearing in mind that techniques and methods used by anthropologists, sociologists and psychologists who normally carry out fieldwork and professional analysis are applied Bautista (2011).

When we need to perform a procedure where the data are held and quantify or somehow perform a statistical analysis to identify the use of numerical magnitudes that can be handled by tools in the field of statistics, we would be working with a quantitative research or methodology. One should keep in mind that this type of research is produced by causes and effects of things, for example, one can have a monetary unit and buy a wheel, but will no longer have that monetary unit.

It is extremely important to be able to maintain a link between the elements of the research problem where it will be characterised by an exponential numerical pattern or similar, bearing in mind that there is a quantitative methodology. This means that the elements that make up the problem must be clearly defined in order to delimit precisely where the problem begins, in which direction it goes and what type of elements make up the problem. Therefore, these techniques are often composed of four concepts called "multi-methods" (multi-methods), "multi-strategy" (multi-strategy) or, to be more clear, "mixed methodology", mixed methods are based on the simultaneous use of quantitative and qualitative methods. For Bryman (2006), the antecedents go back to classical Greece where philosophers like Johnson et al. (2007), taking into account the first uses in both the Social Sciences and Humanities came hand in hand accompanied through Anthropology, Sociology, Science of Sciences.

The influence of triangulation has been realised since the 1950s, when it is recognised that the first mixed methods were applied by Campbell and Fiske (1959); therefore, this concept may be fundamental to the scientific methodological discussion in future decades.

For this investigation, a mixed approach was used to be able to combine and develop most of the stages, taking into account the need to collect information in order to carry out the triangulation. This triangulation aims to find different ways of understanding and interpreting the phenomenon under study.

In this investigation, a mixed approach is carried out, taking into account the need to collect and analyze in addition to linking the data, taking into account that one methodology complements the other and being able to have a more truthful reality about this study and thus be able to respond better in this research. In the quantitative approach the numerical results used with resource to technological tools and the translation of the case study were applied to explain, describe and explore information of vital importance, as it will be analyzed below.

In this research we worked with a sample of 60 students from the 11th grade of the Instituto Politecnico da Graga in Benguela, we applied an instrument in order to validate the variables of this study. In addition, another instrument, the focus group was applied to the teachers who teach the subject of mathematics, which facilitated the collection of data necessary for the planning and elaboration of the methodologies used in the learning of mathematics, both face-to-face and semi-presential. These instruments also served to recognise which were the difficulties in each one of them, recognising the main problems in the learning of mathematics by the students, as well as the methodologies that these teachers applied in the learning of mathematics. The qualitative research found a large number of difficulties in the instruments applied. They used both open and closed questions in each of the instruments used. These data were collected from a mixed approach in the analysis of the diagnosis made in which the problem of study was found.

3.3. Description of the context, participants or population and the period during which the research was conducted.

3.3.1. Characterization of the school.

With the need for society to have a level of training in accordance with the needs of the world around us, the Government of Angola believes in the possibility of building polytechnic institutes in different areas of the country. In Benguela, specifically in the municipality of Graga, the Instituto Medio Politecnico de Benguela (IMPB) was created, with capacity to house more than 1500 students, located north of the municipality of Benguela in the neighborhood of Nossa Senhora da Graga in the Community of zone F. The Institute was created with the creation of the Executive Decree n° 308/06, of 18 of November, the basis of the Education System which was inaugurated on the 13 of July of 2007 with the participation of Her Excellency the Minister of the Assembly of the Republic Ana Dias Lourengo, Excellency, Minister of Education of Angola Dr. Antonio Burity da Silva Neto, next to the Former Governor of Benguela, Dumilde das Chagas Simoes Rangel.

Nature and functioning.

The public institution with a technical and technological component and professional practice in the areas of mechanics, with the cold and air conditioning course; the area of environment with the environmental management course and also the area of electricity with the courses of industrial electronics and automation, energy and electrical installations and renewable energies. In which we work in different shifts in the morning and in the afternoon with regular students and with students of the post-laboral period (night).

Locality and Physical Resources.

It is located in the municipality of Benguela in the same province, forming part of one of the communes in zone F the Instituto Medio Politecnico de Benguela; the neighbourhood is in the community of Nossa Senhora da Graga, as a constituent of zone F.

The Instituto Medio Politecnico, is composed of a directorate, a pedagogical and an administrative office, an amphitheatre (with baths), an office of insertion in working life, a coordination of the after-work regime, two secretaries (general and pedagogical), a room for teachers, sixteen classrooms, eight laboratories, two classrooms, four areas, divided for each one and for the respective referenced courses, changing rooms for administrative staff, students and teachers, a reprography, an internal house for the director, a school canteen, a storage room, a meeting room, a sports field, an internal parking lot, besides the infirmary, library, financial accounting office and a gymnasium.

The amphitheatre and all laboratories and workshops are equipped with video players, CD/DVD players, video projectors, overhead projectors, etc.

3.4. Population.

Population of students in the eleventh grade of the Instituto Politecnico da Graga in Benguela.

Table 3.1.

Population of students in grade 11 of the IPGB

Educational level	Course	N°From students	Total number of students
	Class of ten A	22	22
High School	Class of ten B	23	23
	Class of eleventh C	21	21
TOTAL66			

3.4.1. Sample.

The sample size was 60 students of the eleventh grade of the Instituto Politecnico da Graga in Benguela, to which an instrument was applied to assess the variables of the study. This instrument was applied on a single date to randomly selected students according to the sample fraction for finite populations.

Margin: 5%.

Confidence level: 99%.

Population: 66

Sample size: 60

Where:

n = the sample size. N = the population size.

Population standard deviation which, generally when its value is not available, a constant value of 0.5 is normally used.

Z = Value obtained through the confidence levels. It is a constant value that, if its value is not taken, is taken relative to a 95% confidence level and is equivalent to 1.96 (as usual) or relative to a 99% confidence level and 2.58, a value that remains at the discretion of the researcher.

e = Acceptable limit of sampling error that, generally, when it does not have its value, a value ranging between 1% (0.01) and 9% (0.09) is used, a value that remains at the discretion of the interviewer.

For the sample, both inclusion and exclusion criteria were selected:
a. Inclusion criteria:
- Students from rural areas.
- Students from 15 to 18 years old.
- Students with regular attendance at school.
- Students who wish to participate in the research.

b. Exclusion criteria:
- Students with frequent absences from classes.
- Last students to hand in their questionnaire.
- Students with insufficient information for analysis.

Table 3. 2.
Sample of students

Educational level	Course	N° of students	Total number of students
Bachelor's Degree	Class of ten A	20	20
	Class of ten B	20	20
	Class of eleventh C	20	20
	Total	60	

3.5. Study.

The creation of a virtual learning environment as a support to the teaching process of the subject of mathematics, in which a diagnosis was made to know the needs in the process of learning mathematics; In this work, it is recognized that the investigation reviews the specialized literature related to the virtual learning environments, as well as the different methodological strategies that can help in the teaching-learning process in the mathematics subject, the review of all the bibliography related to the juridical part in which the norms and the legal regulations are analyzed that can support the incorporation of technologies in the educative processes of mathematics. It was also carried out a field investigation to know which is the reality in the process of mathematics learning of the teachers that teach the subject and the use that they make of the technologies in the insertion of the process of mathematics learning in the Polytechnic Institute of the Graga in Benguela, (IPGB).

3.5.1. Data collection techniques and instruments.

For data collection, different techniques were applied that constitute a set of mechanisms, with different means and resources to collect, preserve, analyze and, thus, be able to transmit the data and the investigated phenomena. These techniques and their procedures provided the collection of information in which the researcher will be able to approach and analyse it. Therefore, these techniques are essential to the process of scientific investigation, providing the possibility of organized research.

For this particular investigation, techniques related to field research were applied, meeting the need to collect information directly from the object of study, such as

a. meetings held with the students where the interviews were applied.
b. and the application of questionnaires to students.

These questions were formulated to guide this research, taking into account the system of variables and indicators, so that they serve as a basis both in the construction of the questionnaire, to be able to measure the degree of satisfaction that students have with the pedagogical strategies implemented so far.

Taking into account the objectives of this research, besides the categories of operationalization of the variables, instruments were developed to allow the collection of objective information, where the opinion of the population and the sample subjects was valued; with this information, the questionnaire was elaborated and applied both to teachers and to students and authorities of the institution, these instruments allowed to characterize the real state of the virtual classrooms in the IPGB.

As can be seen in Table 3.3, the questions that were raised in the conceptualization of the variables in the development of the research are presented.

The following table details the techniques and instruments that will be used in the work to be carried out:

Table 3.3.
Data Collection Techniques and Instruments

N°	Techniques	Data collection instruments
1	Search	Survey Inquiry

3.5.2. Information processing.

The information had a quantitative treatment, considering the objectives and the data analysis:

-Tabulary of data.
* Results coding.
* Analysis of results in a logical and reflective manner.
* Interpretation of results with statistical software.

From the quantitative point of view, data analysis was carried out, which was characterized in addition to making comparisons between the information in order to have a degree of reliability of what happens in the classroom, Bartholomew (1990). For this research work, the types of analysis were carried out.

3.5.3. Validation of data collection instruments.

These instruments were used in the data collection of this investigation in which the author used them. They were validated by specialists in the teaching of mathematics with more than 20 years of experience in the sector, having worked in the form of a virtual classroom to support the subject of mathematics, who had scientific publications related to the same theme. All the experts work at the Instituto Superior Politecnico de Benguela (ISPB).

According to the experts, the results obtained were obtained through group, specialized, aggregated, individual or Delphi methods. As part of the group category, the consensus method, already classified in the second category, was named Escobar-Perez and Cuervo-Martinez (2008).

The following table presents the method analysis of the results obtained by the experts and where the instruments are validated in the case of the questionnaire. Table 3.4.

Selection of methodology for expert judgement.

Method	Selection (Yes/No)	Justification
Individual aggregates method	Yes	Taking into account the four specialists who worked at different loading times and used the ideal method, attending to the items of the different instruments. This helps economically, because in reality there is no requirement that the meeting must take place in a certain place. It is a limited method when the specialists cannot exchange opinions and points of view and have to work individually.
Delphi Method	No	These options were considered as another alternative due to the interactive nature of the methodology, which performs a review of the entire group, which disqualifies the Delphi methodology considering that four experts did not have time to work on the interactions of each instrument evaluated.
Nominal Group Technique	No	This alternative is suggested by 8 to 10 experts, in which the scores are registered in the instruments without exchanging opinions. Taking into account that each expert explains his/her reasons to qualify the group. Considering that these actions are disqualified because only four experts were summoned, given the limitations that can be made in a meeting to gather all the experts.
Group consensus method	No	In this method, the experts were gathered only in a single place to perform the task in which the aspects related to the elaboration of the items, recognized by all experts, are estimated; allowing the exchange of information within the groups of experts, reaching a consensus. This method was disqualified due to the experts' conflicts in agreeing on the space and time to validate the instruments.

Note. Adapted from Escobar Perez, J. and Cuervo Martinez, A. (2008). Validez de contenido y juicio de expertos: una aproximacion a suutilization . Avances en Medicion, 6(1), 27-36. Retrieved from
https://www.researchgate.net/publication/302438451_Validez_de_contenido_y_juicio_de_expertos_Una_aproximacion_a_s u_utilizacion

After the analysis performed in the Table, the Individual Aggregates Method was selected for the validation of the instruments for the collection of data for the surveys. Below are the steps that were taken to validate the instruments:

First phase: in this first phase, taking into consideration both the content universe and the specifications and operationalization of the study variables, the four experts collected the information in electronic format.

Second phase: for this second phase, elements such as congruence in the item domain, clarity, tendencies to occur and observances are taken into account, where the experts received the validation instrument in electronic format.

Third stage: in stage 3, the validation instruments collected are analysed:

1. Items that agree 100% favourably with the judges (this means clearly congruent and unbiased) are included in the instrument.

2. Items with a 100% unfavourable match are excluded from the instrument.

The table shows the validation format of the instruments used in the survey.

Table 3.5.

Format used to validate survey instruments.

Instrument name:											
Item	Clarity in writing		Internal Coherence		Response Induction (Vies)		Language appropriate to the level of the informant		Measure what you want		Observances (if an item is deleted or modified for please indicate)
	Yes	No	Yes	No	Yes	No	Yes	No	Yes	No	
1											
2											
3											
...											
N											
General aspects									Yes	No	*****
The instrument contains clear and precise instructions for answering the questionnaire.											
The items enable the research objective to be achieved.											
The items are distributed logically and sequentially.											
The number of items is sufficient to collect the information. If your answer is negative, propose items to add.											
Validity											
APPLICABLE NOT APPLICABLE											
APPLICABLE IN ACCORDANCE WITH THE FOLLOWING REMARKS:											
Name of the expert											
Signature:											
Electronic mail											
Validation completion date											

Note. Adapted from Martin Arribas, M. C. (2004). Diseno y validation de cuestionarios. In Matronas Profesion, 5(17), 23-29 https://enferpro.com/documentos/validacion_cuestionarios.pdf

3. All elements that have been reviewed and judged to be partially compliant will need to be re-validated or replaced.

If we take into account the process of the 3 steps mentioned above, we will see that a consensus was reached among the experts who validated these instruments, such as the written questionnaire.

3.5.4. Purified instruments for data retrieval.

To get these instruments validated by experts, it is necessary to take into account:

Instrument 1 was applied to teachers of mathematics subjects with the aim of collecting information related to the process of teaching and learning mathematics with the support of a virtual environment.

Instrument 2. The instrument was applied to students with the aim of collecting information on how the learning of mathematics is carried out with the support of a virtual learning environment.

3.5.5. Instrument and data analysis.

For instrument 1, the questionnaire was applied to mathematics teachers with the purpose of knowing how they carry out the mathematics teaching-learning process in virtual learning environments.

In relation to Instrument 1 - the questionnaire had 20 to 25 questions.

Purpose: the premise was to collect first hand data from the teachers to know their relationship with the teaching-learning process through the support of virtual learning environments.

Instrument format: this format was composed of 25 questions, taking into account that each one of them provides information about the results of the study. Some details of the section are presented below, such as the questions (1,2,3,4,5,6) which allowed taking into account both theoretical and practical considerations regarding the learning of mathematics and the impact on the teaching and learning process of the subject of mathematics, providing support for the implementation of the virtual learning environment where students worked in a semi-presencial way. These components were based on and reviewed based on the contributions made by the following authors: Gonzalez (2000), De Guzman (1993) and Martinez-Padron (2003), many of these arguments represent attitudes as part of people's subjective knowledge, whose opinions depend on the context in which they were developed, taking into account these opinions individually and collecting both their advances and their failures. For questions 7, 8, 9, 10, 11, 12, 13, 14, 15, 16, 17, it was possible to know the influence of technologies in the classroom, as well as to collect information about whether teachers are prepared to use technologies inside and outside the classroom. All this information was used to make a decision regarding the information that should be integrated in the curricular plan.

Questions (18,19,20,21) allowed us to validate both the teacher-student interaction and the student interaction, in a face-to-face way and in a virtual way; it must be acknowledged that with the application of these forms, collaborative and autonomous learning is achieved in relation to the resolution of the problems raised in class. The questions (22,23,24,25) confirmed that the virtual environment has been a support tool in the subject of mathematics for the students of the eleventh grade of the Instituto Politecnico da Graga in Benguela (IPGB).

In relation to the questions that are part of the items used in this research, many of which were reused, or adapted to the needs of the researcher, Almeida (2010), Carrilho (2006) and Inacio (2006). To carry out the questionnaire, the work carried out by other researchers was taken into consideration where the questions should be simple and also short Hill and Hill (2005), language and clear answers were taken into account, taking into account that they are authentic answers. The pre-investigation was the initial investigation on the use of VLE in Mathematics, being the first validation, which was submitted and analysed by four specialist teachers. If we take Almeida and Freire (2008), we will find that these questions are performed as suggested by the author, through a discussion and a critical approach in their construction, allowing them to be improved.

The questions that predominated were of two types: questions about facts and questions about opinions, attitudes and preferences Freixo (2011). These were mostly closed and multiple choice. Thus giving the respondent the possibility of having several alternatives, being useful when you need information about the variables of study Hill and Hill (2005). In order to profile behaviour and cognitive learning, it is necessary that the teacher is trained in what to teach, how to teach, why to teach, taking into account different ways to achieve their purpose Mill and Pimentel (2010). For Mill and Pimentel (2010) in the world in which we live, ignoring technologies has a very high price and cost, so it is necessary that each teacher is trained in the use of technologies and include them in their daily work for the benefit of the teaching and educational process using methodologies and helping students to improve their performance every day. (See annex 1).

For Instrument 2 - the survey is a twenty-five question format.

Aim: to obtain data from the students in order to know their opinion about the teaching and learning of Mathematics with the support of the implementation of a virtual environment.

Format of the instrument: it was designed with 25 questions, taking into account that it should contain information about the results of this study, considering each session. The questions (1,2,3,4,5,6) highlighted both theoretical and practical information according to the attitudes presented towards the subject of mathematics and to know what the impact was on the teaching and learning process, taking this into account for the assessment. This was carried out taking into account the contributions of the authors Gonzalez (2000), De Guzman (1993) and Martínez-Padron (2003) considering that the information of each subject varies according to the context in which it is being developed, its development is taken into consideration as well as its failures, all through the participants.

The questions (7,8,9,10,11,12,13,14,15,16,17) aimed at evaluating the application of the technological tools in the improvement of the teaching and learning process, besides knowing how the mathematical thinking of the students is developing, their abilities and competences regarding the management of the technologies, it was possible to collect the information for the curricular integration. The questions (18,19,20,21) were able to evaluate the teacher-student interaction in face-to-face and virtual modality, as well as the student interaction in face-to-face and virtual modality, allowing to know the work done in a collaborative way and the autonomy of the students.

The questions (22,23,24,25) showed the contribution of the didactic design of a virtual environment as a tool to support the teaching and learning process of mathematics in students of the eleventh grade of the Instituto Politecnico da Graga in Benguela (IPGB).

This investigation was composed by the contribution of several researchers who, adding different items, composed the questionnaire that is used, Almeida (2010), Carrilho (2006), and Inacio (2006), This work was submitted to the analysis of four specialists and, in its first validation, a pre-survey on the use of VLE in Mathematics was performed, in which it stood out for its clear and understandable language, according to what was suggested by Almeida and Freire (2008).

The questions answered two categories: questions about facts and about opinions, bearing in mind preferences and attitudes Freixo (2011). Closed and multiple choice questions were asked, giving alternatives for selection. In this way, it was possible to collect the necessary information about the variables in this study Hill and Hill (2005). This facilitates their interpretation since they are pre-coded, allowing for greater clarity. Some questions were asked in an open form, delaying the analysis due to the difficulty. For Freixo (2011), these open questions delay the codification of their answers and, sometimes, present problems in their content analysis. Therefore, the use of closed and open questions is useful, according to Hill and Hill (2005), in order to obtain qualitative information, contextualizing the quantitative information with the information provided by other variables. (See Annex 2).

3.5.6. Information processing techniques.

The processing of this information was carried out in several phases, attending to the collection of information according to the objectives established in this quantitative/qualitative research, for which the data were analysed and processed considering the following steps: the tabulation of the data was carried out, the data was coded, the results were collected and analysed logically; in this research the Microsoft Excel tool was also used for the calculation of the statistical processes.

In collecting the information, taking into account the qualitative approach, the following phases of tabulation were included: collection of results, analysis in a logical way using statistical tools related to both quantitative and qualitative research, bearing in mind that each of them complement each other.

3.6. Hypothesis.

This research work does not start from a hypothesis in which it is necessary to demonstrate any results, taking into account the research tasks, and that the research answers each one of them through the results, and that hypotheses can be built as a starting point for new researches concerning virtual learning environments.

3.7. Variables or categories of analysis studied.

According to authors Blanco (2003), Campistrous and Rizo (2003), Hernandez et al. (2006) and Nocedo and Abreu (1984) recognize that the operationalization of variables can be taken into account in the research process, given that the first variable is the didactic design of a virtual learning environment, and the second variable is the process of learning mathematics, both dimensions and subdimensions of these variables are presented in table 3.6; but they can be found in the annexes: (3, 3.1, 3.2, 3.3, 3.3.1, 3.3.2, 3.3.3, 3.3.4, 3.3.5,.3.3.6 and 3.4).

Table 3.6.
Operationalisation of the Teacher Management variable in the VLE.

Dimensions	Subdimensions
Teachers participating in the VLE teaching project (Annex 3.1)	
Students using VLE (Annex 3.2)	
Conception of the didactic design of the VLE (Annex 3.3)	Semantics (Annex 3.3.1)
	Technological (annex 3.3.2)
	Practice (Annex 3.3.3)
	Space (Annex 3.3.4)
	Staff (Annex 3.3.5)
	Management (Annex 3.3.6)
Production process-syperating the VLE (Annex 3.4)	

Note. Adapted from Sanchez Villegas, D. S. (2018). Objetos virtuales de aprendizaje como estrategia didactica de ensenanza aprendizaje en la education superior tecnologica. Ambato-Ecuador.
https://repositorio.uta.edu.ec/bitstream/123456789/28124/1/1804326997-Diego-Sebasti%C3%A1n-S%C3%A1nchez-Villegas.pdf

Taking into account the variable of the learning process of mathematics, being understood from theory and practice, in a receptive way, in which certain processes are performed in solving problems faced by participants in the teaching and learning process, this work here has a different dimension because it transcends the classroom, in view that this learning is supported by the tool of a virtual learning environment for the subject of mathematics. Assuming the dimensions and sub-dimensions of each variable and its results, studies by Bullen (1997), Frias (2008), Gunawardema et al. (1997), Henri (1992), Jarvela and Hakkinen (2002), Scardamalia (2002) and Zhu (1996). Its operationalization is presented in table three point seven (3.7); more details can be found in annexes: (4, 4.1, 4.1.1, 4.1.2, 4.1.2, 4.1.3, 4.1.4, 4.1.5, 4.2, 4.2.1 and 4.2.2).

Table 3.7.
Operationalization of variable: Learning Process in Mathematics.

Dimensions	Subdimensions
Interaction between VLE participants (annex 4.1)	Participatory (annex 4.1.1)
	Interactive (annex 4.1.2)
	Functional (Annex 4.1.3)
	Address (Annex 4.1.4)
	Scope of application (Annex 4.1.5)
VLE participant interactivity (Annex 4.2)	Participatory (annex 4.2.1)
	Functional (Annex 4.2.2)

Note. Adapted from Sanchez Villegas, D. S. (2018). Objetos virtuales de aprendizaje como estrategia didactica de ensenanza aprendizaje en la educacion superior tecnologica. Ambato-Ecuador.
https://repositorio.uta.edu.ec/bitstream/123456789/28124/1/1804326997-Diego-Sebasti%C3%A1n-S%C3%A1nchez-Villegas.pdf

3.8. Description of the data collection for the diagnosis on which the project was based.

3.8.1. Evaluation using the focus group technique.

Spontaneous discussions are held with groups of students about the contents previously selected and guided by the moderator or facilitator. Several authors such as Aigneren (1998), Gibbs (1997), Verdecia (2011) and Williams and Katz (2001) call this technique "focus groups", while Hernandez et al. (2006) call it "focus groups" or "detailed sessions".

The focus group technique should be carried out with a small group of 8 to 15 members, considering that it should have an open and receptive character, in which debates can occur, but always allowing each participant to give their opinion freely without being questioned, allowing the moderator to analyse attitudes and behaviours in order to collect the necessary data.

For the constitution of the focus groups, we considered: the total number of groups, as well as the number of participants, taking into account that they belonged to the IPGB and to the process of teaching and learning mathematics - so we obtained the maximum of 14 members for each group, taking into account the students of the eleventh grade of the Instituto Politecnico da Graga in Benguela (IPGB), which were constituted in focus groups as presented in (Annex 7), with 56 people. A guide of topics was taken into account as shown in Annex 8, which allowed us to know the criteria of each one, both favourable and unfavourable, linked to the study variables. In addition, a workshop was developed for each focus group, where the records and a set of operational criteria were analysed (Appendix 9).

The technique applied in the focus group allowed the author to reconsider some aspects of the didactic design according to the guide for those associated with the operational criteria and the unanimous evaluation, where the collected criteria show that there is a great coincidence in the members regarding the operationalisation criteria of the variable, as well as changes regarding the didactic design of the virtual environment for the subject of mathematics, these decisions contributed to the aforementioned changes.

3.8.2. Evaluation through a pedagogical experiment.

The authors Nocedo and Abreu (1984) consider that the educational teaching process linked to the design of a virtual environment as a support for the subject of mathematics, serve to determine different results such as: firstly, the contribution between the link between the educational teaching process and the virtual learning environment and secondly, the virtual learning environment as an entry level to the study problem in the learning of mathematics.

In the second semester of the academic year 2019, the pedagogical experiment took place until February 2020. The teaching group of the eleventh grade of the Instituto Politecnico da Graga in Benguela (IPGB) was selected for the implementation of this scientific method.

3.8.3. Characterisation of the teaching group and the selected experts. The selected students were part of the sample according to the characterization carried out in the second research task, considering that the same group interview was carried out, as previously done (Annex 5.2). According to the indicators of the dimension "students who use the VLE" (Annex 3.2), the aim was to corroborate the state of the group of the variable of didactic design of a VLE.

The group interview had the results obtained in the sample that was characterised taking into account the amount used in the virtual learning environment in the sample of the virtual learning environment design.

This result of the characterization performed represents a high quantity in the existence of the tools in relation to the virtual learning environment, which have the necessary skills to use them, representing 73% of the limitations for the use of the virtual learning environment in exchange and collaboration. The previously exposed demonstrates how the selected experts present similar characteristics to the initial characterization, reason why the referred selection is considered valid.

3.8.4. Description of the tools or procedures for data analysis. Considering the objectives of the quantitative investigation, the analysis of these data will be carried out considering the following phases:

- Data tabulation.

- Results coding.
- Analysis of the results in a logical and reflective manner.
- Interpretation of results with statistical software.

From the qualitative investigation, an analysis of the collected data was carried out, which were categorized, being possible to synthesize and compare with information already obtained in order to evaluate the reality in the virtual classroom, Bartholomew (1990). For this investigation, two types of analysis will be considered, which will be integrated.

In addition to the conclusions that are obtained from the collection of information in accordance with the objectives set out in this research, always taking into account qualitative research and the analysis of the data processed in the phases: taking into account the tabulation, coding and analysis of the results from a logical perspective, this research will also use tools such as Microsoft Excel for the calculation of statistical processes.

The collection of this information corresponds to the qualitative investigation taking into account the systematization which will be carried out in the following phases: tabulation, codification, analysis of all the results in a logical way, always using tools which facilitate the statistical calculation, both quantitative and qualitative, taking into account the complementation between each one of them.

3.8.5. Description of the creative process undertaken to develop the design.

In the 2019 academic year, we began using the virtual learning environment as a support for the subject of Mathematics in the eleventh grade at the Instituto Politecnico da Graga in Benguela (IPGB).

With the implementation of the virtual learning environment it became necessary the regular presentation of the activities of the subject, not only face-to-face, taking into account that they were related to the same content that was being taught and in the same way contributing to the learning of mathematics. Nevertheless, the final reports of the educational teaching process in the IPGB's teaching department reveal the existence of difficulties in the achievement of these objectives.

To evaluate the contribution of the didactic design of the virtual learning environment in the subject of mathematics, the information that was applied during the 2019 school year to the said sample was collected:

☐ The specialist teachers of the annex 1 in mathematics who participated in the interviews of (annex 5.1); the dimensions were considered: of the teachers who participated in the design of the virtual learning environment, as well as, the production process of the virtual learning environment, belonging to the variable learning process of mathematics.

LI Regarding the student interviews, in (appendix 5.2) and regarding the variable mathematics learning process, we have taken into account indicators such as the students using the virtual learning environment.

It should be taken into account that a documental analysis was also carried out (annex 6)

where information of interest for this investigation is collected, considering the theoretical-methodological references of the following documents:
- The methodological teaching reports of the teaching department of IPGB in the years 20152018.
- The pedagogical reports taking into account the methodological coordination in the academic years 2015-2018.
- The diagnostics carried out in the eleventh grade of the Instituto Politecnico da Graga in Benguela (IPGB) 2015-2018.

For authors like Lopez (2002) and Pinuel (2002), it should be made a content analysis of all activities, tasks, resources and the structure of the virtual learning environment according to the indicators mentioned in this ep^graph (annex 3.3), taking into account the variable virtual learning environment design. This method was also used during the exchange of messages between students and teachers in the virtual learning environment, knowing these indicators of the integration dimension (annex 4.1) and of the interactivity dimension (annex 4.2) of both variables.

For the application of this method, it is necessary to guarantee the evaluation reliability, according to the authors Lopez (2002) and Pinuel (2002), considering that the author was part of the auxiliary team where mathematics teachers with the title of Master participated.

This team performed the evaluation of the content through the proposal of operationalisation of the variables made by the author related to the integration of the teaching and learning processes in the virtual learning environment where the homogeneity, inclusion, usability and mutual inclusion among the indicators were evaluated; the results did not show discrepancies or changes in this proposal.

This team, in order to obtain the reliability of the application of the method, analysed the contents of the virtual learning environment and they were compared by the team in which the author participated. The level of reliability found was between 94% and 97%, being favourable for both indicators evaluated.

In the quantitative and qualitative analysis of the collected data, the following characteristics were obtained in the didactic design of the virtual learning environment for mathematics:

1. Little mastery on the part of teachers in the design of learning as a support for mathematics with a virtual learning environment; little use of the processes of production of the virtual learning environment in the pedagogical improvement of the mathematics subject.

a) In the investigations carried out, 22.7% of the teachers and 14.3% of the specialists received pedagogical training in the work of the virtual learning environment. Out of 12.4% of the teachers and 7.2% of the experts who participated in the virtual learning environment project, only 14.8% achieved 3 of the mentioned components. Likewise, less than 12% of the teachers and none of the experts knew about the production of the virtual learning environment, nor participated or confirmed their knowledge in the integration of the virtual learning environment.

b) These interviews with groups of teachers were not able to provide a definition of virtual learning environments where very few were able to express what their components were; many explained that they were not used within the pedagogical model of the mathematics discipline. Many of those who participated in the didactic design of the virtual environment and the configuration of these activities expressed that it was the first time that they had participated in a process similar to this one, because this type of experience was not highly valued and they did not know the results that it could bring to the process of teaching and learning mathematics.

c) In the document analysis, according to the reports of the IPGB department, there is a need for better preparation in the use of educational technologies.

2. On the insufficiencies of the didactic design of the virtual learning environment in relation to the anaKtic programme of the mathematics subject.

a) The interviews granted to the group of teachers representing 67% expressed that the didactic design of the virtual learning environment was centred on the designation of materials of knowledge systems, and that it did not allow the practical activities that are carried out in the face-to-face system, which can cause a poor learning of mathematics.

b) In relation to the documental analysis reports of the IPGB department, it is possible to know that the didactic design of the virtual learning environment, that is a repository of didactic materials for each one of the programme units, most of the activities are not used by all the students and are only used by some of them.

c) With the analysis of the activities of the contents in the virtual learning environment for a 44%, it can be clearly seen systematically that they are related to the AI problems solved with the objectives studied and achieved, it can also be verified that the contents were used through the organization methods, which represents that they are part of the expected results. This demonstrates that sometimes, when it was necessary to work in face-to-face classes, this type of activity had to be carried out individually in order to attend to differentiated attentions, and often took a long time, while now not only is a differentiated attention obtained, but also an interaction where the students participate, exchange opinions among themselves, and the teacher becomes a facilitator.

3. The use of the virtual learning environment as a support for the mathematics subject in the teaching and learning process was very well accepted by both students and teachers, given the facilities it offers as a support for this subject. The specialists' evaluation of their satisfaction with the didactic design can also be considered positive.

a) All the experts who used the virtual learning environment system gave a positive opinion, 2.3% of the expert teachers who used the virtual environment expressed that they were dissatisfied with the design of the virtual learning environment.

b) From the teachers who were interviewed, 82% recognised the virtual learning environment as a means of integration in the teaching-learning process that reduces the space-time barriers, but 77% considered that the virtual learning environment should be redesigned so that it could reach 100% integration with the projects proposed in the subject.

c) It can be seen in the interviews with the students that 65% expressed unfavourable criteria of the virtual learning environment many times because they were not prepared and other times because they did not understand the tool due to the lack of ability to work with these types of tools, but at the end of this investigation they managed to overcome.

4. It is acknowledged that in the context where the research was carried out there is a low application of technologies in their integration with the educational teaching process, becoming a scenario in which technology is sometimes rejected.

a) In interviews conducted with teachers, 17% were able to explain how they carry out activities in virtual learning environments, tasks and how they can use certain resources belonging to this tool.

b) 69% of the interviewed students recognized that the use of the tools used in the virtual learning environment not only facilitated their understanding of the content, but also helped them to communicate and help other students to exchange different points of view regarding problem solving and, at the same time, also allowed them to communicate with the teacher who not only provided content, but also helped them to solve different problems. Although it must also be acknowledged that most of the students ignored the use of tools such as the glossary, blog, wiki, used in the virtual environment.

c) It was possible to verify in the content analysis of the virtual learning environment structure that only 43% of both the tools and services used responded to the anaKtic programme of the subject.

d) In the document analysis collected by the IPGB department it was possible to collect information that even when technology is used, proprietary tools are used, which makes its use difficult due to the need of paid licenses, when in reality there are free format tools that offer the same services.

5. The didactic project of a virtual learning environment, its objectives and the interaction with the teaching-learning process system.

a) The content analysis of the virtual learning environment structure demonstrated the possibility for each teacher to design their own environment according to their projection and the content they will offer in that virtual learning environment, taking into account to whom that content is directed.

b) In the content analysis of the virtual learning environment, it was possible to perceive how the practical activities were developed where there was an interaction between students and students and teachers-students, where ideas were exchanged and collaborative work was carried out.

c) In this content analysis of the activities, the tasks have played a very important role, since in these learning environment spaces an interaction is achieved where the student can carry out a self-evaluation to critically recognise where his weakest aspects are in relation to learning and be able to reinforce this learning to obtain a better result allowing an individual and collaborative work of the participants.

6. The management of the virtual learning environment taking into consideration the management of this environment and the mechanisms that can support this system.

a) In the analysis of the activities of both the tasks and the structures of the virtual environment, 6.2% of the teachers had access to the edition of the course as such, which made it difficult to implement it in the didactic design, having to adapt it to the particular characteristics of each teaching group in each student. In the IPGB there are no other components in the VLE with the function of technological or pedagogical assistance to the participants.

b) As for the analysis of the contents of the virtual learning environment, only 12.3% of the resources presented in this project, taking into account that these platforms are found on the Internet, often the maintenance of these environments is a little difficult when the networks do not have enough bandwidth to be able to work on them and several hours of work are required, sometimes making maintenance difficult.

7. An insufficient definition was made in the design of the virtual learning environment, taking into account the development of this software, which was necessary for the management of the technology-based teaching and learning process.

a) Therefore, in the content analysis of the virtual learning environment structure, the roles to be played by each of the participants were sometimes affected in this case, often the participating teachers who are facilitators and the students 62 who must perform different types of tasks in this virtual learning environment.

b) Despite what was explained before, it was possible to define the teachers' work taking into account 3 well defined roles, the first of them as a main teacher, the second as an editor teacher and 1/3 as an editor rights teacher where the first occupied represent 2%, 3% and 6.2% respectively, leaving 91.5% of the teachers without the possibility of controlling the students' work in the environment with the possibility that the teachers who are participating in the project can control the activities of the students who are participating in the virtual learning environment.

8. Taking into account the diffusion and not the integration of these processes, the experts agree that the virtual learning environment in the subject of mathematics has been able to support the learning process of this subject.

a) 76.2% used the virtual learning environment; only 94.6% per week, and 71.7% of the experts did not use it. 89.3% of the teachers stated that they used it for the knowledge of the others.

b) The interviews conducted to the 18.7% groups of teachers recognise that there are different ways of designing virtual learning environments in support of learning processes.

9. The low interaction of the participants in the virtual environment, which are not in a chain and do not fulfil social-technical functions in a horizontal or partial sense.

a) In the analysis of the content of the messages that participated in the virtual learning environment, it was obtained a frequency of emission of messages per participant of 4,2%

per day for students and for teachers 5,4% as the participants were integrated in the tasks. It became 41,3% in 32 school weeks, taking into account that more than 71% of the content had a social or technical function related to the approaches of this mathematical subject and that they were oriented in the usual learning environment.

b) In the analysis of the content of the messages of the participants in the virtual learning environment, 17.6% were linked, 79.4% in a horizontal way, to the content that was being taught in the virtual learning environment.

c) 91.4% of the messages referred to the results and experiences to support the project of the mathematics virtual learning environment, these contents were elaborated in messages by the participants of this environment.

10. The interaction in the function of consulting the predominant contents and the contribution of this work in a collective way in learning with others.

a) In the content analysis of the participants in the virtual learning environment, a contact frequency of 7.1% of times a day was reached for the students and 5.8% of times a day for the teachers, taking into account that the participation in the interaction was 1293 in 32 teaching weeks.

b) In the analysis of the contents of the participants in the virtual learning environment, 21.4% had contact with the objectives of the virtual environment from the contribution of the contents therein.

It was not possible to identify a level of correspondence in the didactic strategy elaborated in the subject and the teaching-learning process in the semi-attendance modality. This didactic design of the virtual learning environment needs to be more studied and mainly modified in the school curriculum, having in mind that the technologies must be part of it for what these types of tools represent nowadays in the school environment. It was shown that few teachers have the necessary knowledge to be able to apply the technologies in the teaching and learning processes, reason why it is necessary a previous preparation to acquire the necessary competences and not reject them by ignorance.

3.9. Ethical considerations.

We have taken into account the ethical printpoints that justify this investigation in accordance with international standards and national resolution 008430/93.

In this research, we have taken into consideration the opinion of the legal representative of the institution, according to the research conducted; containing the feeling of all participants and the approval of the research ethics committee in this institution.

CHAPTER 4
DIDACTIC DESIGN OF A VIRTUAL LEARNING ENVIRONMENT AS A SUPPORT TOOL FOR THE SUBJECT OF MATHEMATICS FOR STUDENTS OF THE ELEVENTH GRADE OF THE POLYTECHNIC INSTITUTE OF GRAQA IN BENGUELA.

Chapter four responds to the general objective, taking into account the didactic design of a virtual learning environment as a support tool for the subject of mathematics for students of the eleventh grade of the Instituto Politecnico da Gra?a in Benguela. The contents were selected in accordance with the curriculum of the mathematics subject, and the evaluation was used in accordance with this proposal.

This platform included not only the tasks that were carried out according to the content of each unit, but it was also possible to work with tools such as the virtual forum that allowed the debate and exchange of ideas between students who contributed new ideas and if they had any needs, they could also ask the teacher for an exchange; moreover, it was also possible to work in a collaborative way with a more critical thinking and contributing to the knowledge of mathematics, as can be seen in the following link:

https://rbcc.milaulas.com

4.1. The role of the teacher and the student in virtual learning environments.

The possibilities that technologies offer on an educational level are immense if, for example, one considers the possibilities related to the creation of a course, one could be talking about a tool such as EMULE that allows the creation of courses and administers not only one subject course, but also all the courses of an institution, which facilitates giving the student a tool with which he can be a participant in the construction of his own knowledge of the subject. The introduction of technology at the Polytechnic Institute of Gra?a in Benguela has made it possible to create a virtual learning environment to support the subject of mathematics, making it easier for the teacher to manage the course and to use it as a support for the subject of mathematics for a better understanding on the part of the students; this has made it possible to have systematic control over the activities of the students not only in problem solving but also in their interactions, noting how the knowledge of the students has progressed little by little in this subject. For the student, it has been the possibility of having new alternatives of how to learn to interact in an environment in which control 65 tools that they normally use in their daily social life, using the didactic materials found on the virtual learning environment platform.

The idea of creating a course in a virtual environment of mathematics was not to create the same model that is found in the face-to-face mode, but to use several elements that need to be reused, but taking into account that the platform offers several tools in which it is possible to deepen several contents and that can serve as support to the students so that they can have a better understanding of certain contents. In the work with the virtual platform, we take into account other variants that count in the face-to-face mode, but not as in the virtual platform, as in the case of the communication that has a relevant value, taking into account that it is through it that the student-student and teacher-student interaction is established in a way that goes beyond the face-to-face. In this environment, technologies are the mediation tools between the teacher who is the content facilitator and the student who is not only there to learn, but also to build his own learning. It is necessary to keep in mind that technologies not only represent the linking of content, but also provide spaces where there are opportunities to discuss and interact in order to enrich learning. By including technologies as educational resources, not only content and activity proposals are addressed for the strengthening of student learning, but also for learning to work individually and collaboratively with other students in a group, So that they can share critical thoughts and also be able to make contributions where other less advanced students can also participate and learn within the most important themes the student will have the possibility to make their own evaluation and recognize where they are and what their mistakes are and be able to correct them and improve their own learning, maintaining constant communication with the teacher who becomes the facilitator of learning for these students.

Professor.

The collective of teachers of the mathematics discipline will be responsible for the development and implementation of this course. To provide multidisciplinary assistance and pedagogical services capable of providing learning strategies and didactic methodologies, all within the department of mathematics.

The tutor of the learning environment course before starting the activities of the process, should make known the possibilities and limitations of this platform in which an analysis of the different tools available in the different levels can be done, such as

- Management and administration;
- Communication;
- Content development;
- Development of interactive materials;
- Collaboration;
- Evaluation; and
- Monitoring and customisation.

In virtual learning environments, teachers have two types of tasks to perform, being that the teacher has the possibility to edit the course design and also the possibility to follow the student's performance during the course, taking into account that in this type of course the design of the activities to propose is indispensable so that the contents can be performed with the required adequacy.

Among the main fungoes in Mudanga, one can find the following:
- The creation of the activities of all resources with the tools they provide.
- The planning of when activities should be handed in by students.
- Control of the evaluation of these activities.
- If you can appoint tutor teachers.
- It facilitates the work of tutors by providing support and monitoring the activities carried out by students.
- It facilitates the creation of groups of students accompanied by their tutor.
- Allows the enrolment or cancellation of students on the course.
- It is possible to have the students who are registered in the course followed up by a tutor.

Its fungoes include the following:

The aim is to get to know the student from the personal variant, how he relates to his environment, how familiar he is with technologies and how well he does in the course.
- To get students motivated to participate actively in the course.
- Follow up with students who are not participating in the course, taking initiatives to obtain their insertion in the course activities.
- To achieve an interaction between student and tutor where all the doubts that may arise during the activities that are being carried out in the course can be clarified.
- To follow up all the students participating in the course to know how they are performing in the activities proposed in the course and to give support to all the students who need it.
- You should qualify all the sites that students respond to.

Students.

For the students in virtual learning environments, the role they will play will be totally different, taking into account that they will face technologies that they will have to know how to use to be able to reach the content of the activities proposed by the teacher. Once the students dominate these technologies, the options for their performance will increase taking into account that it will facilitate teamwork where they can collaborate and change their ideas to give contributions among the students themselves and also ask for help or exchange ideas with the teacher, in the same way that they could also do in face-to-face classes; but in this new modality they will discover that it can be done not only in the institution, this type of interaction facilitates, no matter where they are, the distance they are, they will be able to carry out all these options.In addition, they will be able to carry out self-evaluations in order to know what their weaknesses have been, recognise them, refine them and improve them. This will undoubtedly give a new twist to the student's own performance in the construction of their learning.

Among the activities that the student must carry out are:
- To have the possibility to actively participate in the whole course taking into account the activities of the discussion forum, tutorials, chat.
- Issue the activities with the group of students where they can collaborate and exchange ideas including the teacher or tutor.
- Comply with the deadline that represents the completion of activities.
- To be able to carry out self-evaluations in order to determine their weaknesses and strengths.
- To help other students in carrying out the proposed activities, developing a critical thought in the evaluation of the work of their own colleagues, making contributions to these contents.
- To have contact with bibliography which will facilitate the acquisition of significant knowledge and the possibility of putting it into practice in their daily lives.
- To refine the technological skills applied to education.
- Develop teamwork skills.
- And to strengthen communication skills, as well as work organisation and planning.

It means that the student goes from being a passive student in his own learning to an active student where he participates in the construction of his own learning, taking into account the tools of the virtual learning environments, fundamental in the communication, collaboration and interaction of these activities in the course.

The **motivation** is a key element in which the student will seek to discover a new knowledge that he/she will be able to interact through the networks in a context where he/she will be outside the classroom, will be able to listen and exchange ideas, will be able to criticise and be criticised in the way he/she carries out certain activities, will be able to critically evaluate and accept his/her weaknesses and potentialities this new type of learning creates in the student a new motivation in the learning of mathematics.

Within the possibilities offered by virtual learning environments, it is important to give students self-management autonomy, which will facilitate not only the competences they should develop to reach the proposed objectives in the contents and proposed activities, but will also allow them to study, to develop organisational competences in the carrying out of activities, time planning to fulfil such activities, giving them autonomy in the way they should carry out their learning process and in the resources they will use to fulfil the respective activities.

Reaching this level will allow the student to acquire a degree of responsibility and maturity to be able to respond to the demands of this type of semi-attendance courses. Besides, it is necessary to bear in mind that studying at a distance is as difficult as studying in person, it requires as much time and effort as studying in person. In distance learning you must use tools and technologies connected to Internet. Most of its content, activities and resources will be in digital format. Taking this into account, it must be considered that the person who decides to study in virtual learning environments must know how to regulate the time he or she dedicates to these activities, where he or she will be separated from the family and recreational activities that he or she is normally used to due to age; this without a doubt, reactions of rejection to this type of study can happen, so once he or she adapts to this work dynamics, he or she will reach a higher level of maturity and responsibility for their own learning.

1.1.1. Mathematical competences that are enhanced by the use of VLE.

The Angolan Ministry of Education recognizes that it is possible to talk about curricular plans for significant learning in learning by competences. There are precise proposals about the general competencies that students must develop in the educational process, among which are: interpretative competence, anaphatic competence and proportional competence.

Regarding the mathematical competences, the need for logical reasoning in the face of problem solving where the intention is to improve the performance in the study of mathematics. It is intended that these skills are achieved spontaneously, but support tools are needed where virtual learning environments can serve as a stimulus in these problem situations and facilitate the achievement of these skills by students.

In order to achieve the above, it is intended that in the virtual learning environment methodological strategies are developed to improve the learning processes in the area of mathematics, which are requirements of the Ministry of Education of Angola to achieve general skills and mathematical skills in a specific way.

In the study of Mathematics, the support provided by the technologies to its learning is recognised, since they offer multiple options in the work of the different areas of Mathematics in which it facilitates a better understanding by the students of some of these units, facilitating the pedagogic work that the teacher carries out. Allowing the students, through personal computers, to carry out individualized work, using different tools that allow and facilitate this type of study, which does not mean that they can not collaborate in teams with other students, since these same tools allow teamwork maintaining an interaction between the other participants, as well as individual performance, also maintaining an interaction with the teacher, providing the teacher with the possibility of monitoring the student's own performance and the difficulties or strengths that each one of them presents in the development of the proposed activities within the tool of virtual learning environment.

This is relevant in view of the possibility that it presents not only the use of tools in the educational world that today society uses for its own communication and interaction in communication, information and its own learning outside the educational institutions, which strengthens the institution and the curriculum of the mathematics subject with the use of these technologies already used by the students themselves in their daily life.

Which acquires dimensions far beyond those that had been anticipated since the student acquires skills such as: analysing, synthesising, searching for information about contents related to the proposed activities, making analogies of different situations, developing critical thinking not only of their activities in problem solving, but also in the solving of problems that other students carry out when there is an exchange of ideas; They have also developed critical thinking in the assimilation of other students' criticisms of their own results, they have reacted quantitatively and obtained qualitative information from a quantitative database, they have been able to develop strategies in solving the problems they face, and they have improved their mathematical language, bearing in mind that in order to communicate and interact they need to do so accurately.

For the spedific competences one finds a student capable of facing problems of daily life in which they demonstrate what they have learnt in solving mathematical problems. To achieve this, the use of calculation tools, to see errors when solving problems, to be able to revise the results, detect and correct errors, to be able to interpret mathematical problems, to develop skills with computational tools that allow statistics to be made, and to analyse and prove the data provided in certain situations or certain problems. In this case, it would be Applied Mathematics to problems in the area of the Environment as a course of Environmental Management and the area of Electricity as courses of Renewable Energies and Energy and Electrical Installations of the Polytechnic Institute of Graga in Benguela, although the statements only mention problems or real systems without delay, not to make them longer.

4.2. Training model used, type of institution, type of course.

For Tejada (1999a and 1999b), there are model schemes developed from a planning, a programme or a course that require diagrams to guide the process and there are also other proposals that can be considered.

4.2.1. *Collaborative models.*

The models in the teaching and learning process require collaborative work and interaction of the participants. Bearing in mind that not all participants perform the same functions. In the reality of the educational model it will be noticed that this will depend on the objectives, the needs, the strategies of the contents that are proposed for a certain course. In this programming, it will be composed mainly by teachers who will be responsible for the planning, execution and evaluation, considering the educational institutional lines. It will be based on the educational model centred in the students and in the collaborative groups that will work in the virtual learning environment. Where students will be responsible for their own learning process, carrying out investigations, working with the contents and proposed activities through this technological tool, allowing them to develop their own learning process, where they will reach the educational objectives.

The Caldwell and Spinks (1986) model is an example of this approach.

Figure 4.1.
Caldwell and Spinks' collaborative CSM model (1986)

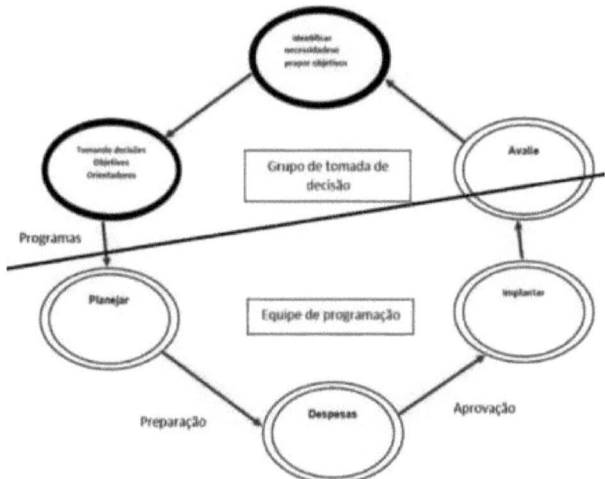

Note: This figure refers to the student-centred educational model and collaborative groups. Prepared by CALDWELL, B.J. and SPINKS, J.M. (1986): Ecole de planification de l'elaboration des politiquesEfficacite. Departement de l'education Tasmanie, Hobart. https://files.eric.ed.gov/fulltext/ED352713.pdf

All the members of this activity will be protagonists of the educational training model, where they will participate in the strategic needs and contents of the course; being the teachers who will make the decisions of that planning and execution and evaluation within the institutional policies. The collaborative work of the students will be taken into account, in which activities will be carried out where they will exchange ideas, experiences, contributions, in the fulfilment of the same. All this will be done by means of the virtual learning environment tool which will facilitate this work.

This model is represented in the form of six phases, differentiated by a group that makes decisions and another that carries out the program. The decision making group will be responsible for the needs and the fulfilment of the objectives according to the needs proposed in the course. The program group will be responsible for the planning and the strategy to be developed in the program. The execution of the two groups will have to be collaborative in order to guarantee that this decision-making of both will achieve the realization and satisfaction of the needs proposed from the beginning of this activity. Starting with a new cycle of unsatisfied needs and restarting this sequence of curricular design and development.

4.2.2. *Type of institution.*

The school is called Instituto Medio Politecnico de Benguela do Ensino Secundario Profisional and is public in nature and has a technical-technological training component and professional practice in different areas, from mechanics including the cold and air conditioning course; the environmental area with the environmental management course; and the electricity area with the industrial electronics and automation, energy and electrical installations and renewable energies courses.

4.2.3. *Type of course.*

The socio-economic development in Angola is mainly linked to two essential elements highly interrelated; on the one hand, the improvement of the base of existing technical material, with the application and use of science and the creation of efficient, advanced and sustainable systems and, on the other hand, the development of its productive forces from competent professionals in each specialty. In this important relationship the Professional-Technical Training (FTP) plays and will play an irreplaceable role, as a broad context of the worker's training process, fundamentally at medium level: Qualified Workers and Medium Technicians.

For this research we worked with the courses:

Environmental Management: The performance of its functions should be prepared for the introduction and integration of environmental dimensions processed for the satisfaction of socio-environmental attributes to train professionals with environmental culture and contribute to the management of material resources in schools and in certain territories, as environmental management of sustainable economic and social development.

Energy and Electrical Installations: This course aims to develop professionals to carry out electrical installation tasks in all types of buildings, whether residential or industrial, as well as maintenance and repair of electrical circuits at a professional level. With these materials, the student will be able to take a broad tour of fundamental aspects such as symbols, the degree of electrification of a building, the electrical circuits or the partial and total loads that each installation can support.

Renewable Energies: In this course, he coordinates the erection, commissioning and operation and maintenance management of wind farms and facilities, promotes installations, develops projects, manages and carries out the erection and maintenance of solar photovoltaic facilities and manages and supervises the erection and maintenance and performs first level operation and maintenance on electrical substations.

4.3. The Profile of the target student and their learning needs.

Vocational Technical Medium Education (EMTP) constitutes a valid alternative for students who want to acquire the first tools that will allow them to develop in the work environment, which shows the social impact that a secondary school student has when making the decision to choose a technical career, as it implies preparing for work from an early age, opening the way to a new world that will serve as the basis of his experience that will allow him to continue to grow as a person and as a worker.

The students who are being trained in the Professional Technical program in the area of Environment as the Environmental Management course and in the area of Electricity as the courses of Energy and Electrical Installations and Renewable Energies; the goals and objectives of the Professional Technical Education are the following:

a) To promote the articulation and integration of Technical Education at secondary level and professional training.

b) To establish links between the institutes and programmes of Technical Vocational Education with the fields of science, technology, production, work, art and sport.

c) To elaborate curricular proposals according to the needs and potentialities of the socio-economic, regional, provincial and national cultural context, integrated with scientific, technological, development and innovation processes.

d) To promote the articulation of the productive sector with Professional and Technical Education.

e) To promote the recognition and certification of knowledge and skills, as well as the voluntary reinsertion in formal education and the continuity of regular studies at different levels and modalities.

f) To develop skills with value and significance in employment, which facilitate professional insertion and promotion.

g) Consolidate the training of workers as citizens by providing them with education in accordance with socially recognised standards within the framework of continuing education.

h) To structure a comprehensive, hierarchical and harmonious provincial policy linked to national policy.

i) To generate mechanisms, instruments and procedures for the organisation and regulation of technical professional education.

j) To improve and strengthen institutes and programs of secondary, higher and technical education and in the field of professional training, through plans and programs that promote the continuous improvement of the quality of institutions.

k) To develop opportunities for specific training in the chosen profession or occupation and for professional practice in the chosen professional area.

Professional training has as its specific objectives:

l) To prepare, update and develop the capacities of people for work, through processes that guarantee the acquisition of scientific and technological knowledge and the mastery of basic, professional and social competences required by one or several professions defined in a broad occupational area, with insertion in the economic-productive sphere.

m) To promote the recognition and certification of knowledge and skills, as well as the voluntary reintegration into formal education and the continuation of regular studies at the different levels and modalities of the education system.

n) To guarantee social promotion, raising the level of qualification of the population, providing opportunities for personal, professional and community growth.

o) To consolidate worker training as a citizen by providing education for and in the workplace according to socially recognised standards within the framework of lifelong learning.

p) To implement body and motor practices which act in a compensatory manner on postural habits associated with the world of work.

Graduates of IMPB's Environmental Management Technician career will have the necessary skills to develop in the environmental management of public and private institutes and bodies, by implementing management plans and optimizing resources, complying with existing regulations and protocols. They also receive significant mathematical training, have research skills and the ability to generate specialized knowledge in environmental systems, natural resource management, and the management of public policies and programs related to the environment; and can operate with a high degree of competitiveness in all activities that require the use of systematic information for decision-making, project design or program implementation in the academic and public, private and social sectors.

After completing their studies, the Renewable Energy Technician of IMPB will be able to assemble and install, maintain, inspect and design small and medium scale Renewable Energy projects, solar photovoltaic, solar thermal and wind power, for the productive activities of different industrial sectors. Moreover, they receive important mathematical and electrical training; they are prepared to develop business projects associated with Renewable Energies, photovoltaic solar, thermal solar and wind power; all of this with an emphasis on the development of practical applications in the context of their speciality. In its training it promotes the development of integral people, aware of their transcendent dignity, and committed to people and society, integrating ethics, values and the development of the necessary competences for today's world, and this is how it is valued. Moreover, the ability to work in teams, to solve problems, as well as the ability to generate innovative ideas, to learn and to constantly update.

The IMPB Energy and Electrical Installations Technician is capable of carrying out assembly and commissioning tasks of electrical systems and installations, repair of industrial equipment and installations and also carrying out tasks of monitoring electronic devices, diagnosis and failure analysis of equipment and systems involved in the company in order to maintain the operation of the plant in accordance with the quality and safety standards in force. Receives important mathematical training, is prepared to carry out industrial lighting projects and normal consumption, giving an option of comfort to the end user. Being capable of dealing with situations and solving problems related to the scope of the organisation in which they operate, applying technical skills in the area and adaptability, communication, effective collaboration with the work team and personal and group commitment.

4.4. Methodology implemented for the development of pedagogical practice.
4.4.1. *Implementation of pedagogical practice.*

Directed appropriately to the 66 students of Mathematics of the 11th grade of the Instituto Politecnico da Graga in Benguela, the teaching and learning process was a fact, 27 classes were held in the virtual platform Moodle that contains the course of mathematics on the topics: Successes; Introduction to differential calculus I; Geometry in space; the overview of the respective topics is in (annex 12); with the respective learning guides, documents, educational videos, and respective evaluations; which are evaluated through the learning platform of these topics with evaluations both through the Moodle virtual platform and directly with the students.

4.4.2. Evaluation of pedagogical practice.

The evaluation was made with the help of the platform results with its evaluations of the learning topics with those involved in the pedagogical practice, i.e. done with the Environment area, with the Environmental Management course and the Electricity area with the Renewable Energies and Energy and Electrical Installations courses of the same school on the same subject, but without the help of a virtual platform.

It should be noted here that an investigation has been carried out directed to all the participants of the pedagogical practice in such a way that it leads to a better visualisation of the usefulness and advantages of this didactic tool in the formation of the students.

4.4.3. Methodology implemented for teaching practice.

This work was carried out under the following parameters:

1. The location is: Instituto Politecnico da Graga in Benguela, located in the Munidpio of Benguela.
2. Area of: Mathematics.
3. The teaching-learning topics are: Successes; Introduction to differential calculus I; Geometry in space.
4. Aimed at students in the eleventh grade.
5. Teaching strategy: Based on the development of new technologies
6. Time: Academic Year 2019.
7. Internet in the institution.
8. A virtual platform: Moodle.
9. Moodle server mounted on the institution's computer.
10. Course developed on Moodle platform with the themes: Successes; Introduction to differential calculus I; Geometry in space.
11. Analysis of results, conclusions, recommendations and future work.

Initially, a moodle server was installed in a computer of the institution with Internet inside the school to broadcast the signal to all students involved in the pedagogical practice.

Three classes were formed from the eleventh grade to attend the teaching-learning of Successes; Introduction to differential calculus I; Geometry in space; teaching-learning of the mathematics course supported by a virtual moodle platform.

Later, the practice with the results in relation to student learning in relation to the subjects Successes; Introduction to differential calculus I; Geometry in space; making an analysis of the strengths and weaknesses for both the results obtained with the teaching groups supported on the moodle platform.

The analysis of the results carried out was of vital importance for the project, allowing to prove the advantages of an education supported in the development of new technologies such as a virtual platform or, leaves the information of a formation in the educational practice showing improvement in the learning acquired by the students.

The final objective of the course as a teaching tool is to determine the objectives for teaching, the central tendency measures, by working the concepts in solving problem situations and at the same time in structuring the contents so that they have an impact on the real practice of the students' daily life.

4.4.4. Activities undertaken.

The activities were developed, based on a virtual moodle platform, for the conception of a course in the area of mathematics, specifically in Successes; Introduction to Differential Calculus I; Geometry in space, previously with the selection of the contents, selection of the population, acquisition and assembly of the virtual platform according to the list below:

1. Determination of the groups of students with whom the practice of the same media training following the traditional model was carried out.

2. The software selected for the virtual course was moodle because it is useful and free. Here, it was supported with: (https://rbcc.milaulas.com/).

3. Induction of the target population (students of the 11th grade) in relation to the management of the virtual moodle platform.

4. Content design: 3 concepts were collected in the form of didactic statistics, which were selected according to the programming in the area of mathematics according to the standards of basic skills issued by the Ministry of National Education.

5. For the development of the contents the following programmes were used:
 - Word-PDF as a word processor for the creation of documents.
 - PowerPoint as a tool for creating presentations.

4.4.5 Main page of the virtual course on the moodle platform.
Introduction of the Virtual Learning Environment

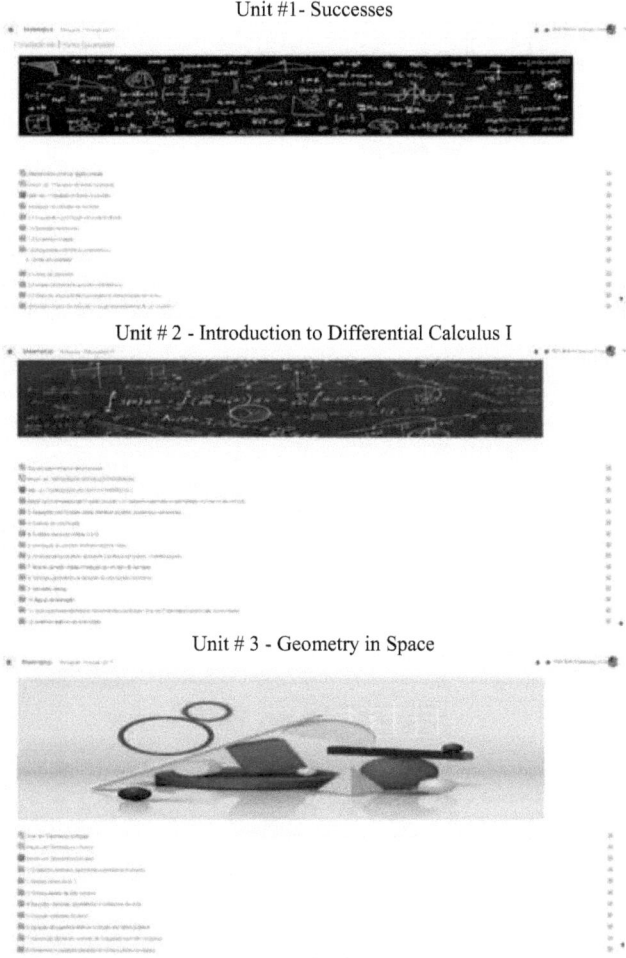

In addition, the following actions were summarised in relation to the virtual course:
1. User registration.
2. Inducement to enter the platform.
3. Conceptualization of the topics from a virtual mathematics course installed on the moodle virtual platform.
5. Motivation of students to enter as an extra-class activity and curricular support to the various themes of the mathematics course on the moodle platform.
6. Development of Chat in virtual classroom for the discussion of the themes of each unit.
7. Access to the virtual forum for deliberation in digital format and the respective uploading of answers to the platform.
8. Virtual presentation of videos for further discussion in class and submission in the notebook of the respective summaries on the theme of the videos watched.
9. Application and deepening of concepts developed in the classroom through videos and

simulations on the virtual platform.

10. Wiki review and support.

4.4.6. Activity summary table.

The following tables compile the various aspects related to the activities proposed during the development of the pedagogical practice, taking into account those responsible, the objectives, their description, evidence, achievements and the role of the teacher as a guide and facilitator of the process.

Table 4.1.
Activities related to pedagogical practice.

Activities	Responsible	Objectives	Description	Evidence	Achievements
Induction input for platform	Teacher	Orienting students to enable them to access the platform.	A short explanation of how to access the platform, distribution of content and tools which presents the Virtual Platform.	Records on the platform.	Permanent Student input to the platform.

Table 4.2.
Activities related to pedagogical practice - thematic contents.

Activities	Responsible	Objectives	Description	Evidence	Accomplishments
thematic contents:	Teacher: Design, construction and material implementation. Students: access to each of the planned activities.	To contribute to the understanding of the application of the different concepts in practical life. Through a course installed on the virtual moodle platform.	The students visit the virtual course on the platform where they can get to know and find the themes a develop academically and following the tutor's guidance and course guides. If take place group discussions of the video lessons. Comments on the themes of the course (class). Access to links of other paginas web (extra class). Reviewed by knowledge. Assessments (inside and outside the classroom. Solve concerns in class also for the students; and the teacher socialises the rest of the students.	Visits to the course for the development of themes cited. Solved tasks on digital form and upload to the platform. I participate in the forums. Resolution of the online assessments.	Develops ICT skills. Gereum Learning Autonomous, collaborative and meaningful.
Successes	Teacher: Design, construction and material implementation. Students: access to each of the planned activities.	-Definition of succession. - Visit different ways of defining a succession. - Calculate terms of a succession. - Investigating whether or not a given number is	The students visit the virtual course on the platform where they can get to know and find the themes a develop academically	Visits to the course for the development of themes cited. Solved tasks on digital form and	Develops ICT skills. Gereum autonomous, collaborative learning

		term of a succession. - Estudara monotony of a succession. - Calculate, when exist, the sets of minorantese majorants of a set. - Check that a succession and limited. - Set progress arithmetic and geometric. - Calculate the ratio of a progression. - Write the term general of a progression. - Calculate the sum of n consecutive terms of a progression. - Operating in the set [-",+"]. - Proving that a succession and a infinitesimal and its inverse and an infinitely large one. - Define the limit of a succession. -Raising indeterminacies of type 1^a да.	and following the guidance from tutoring guides about the course. If take place discussions at video clusters - classes. Comments of the themes from course (class). Access to links from other paginasweb (extra class). Reviewed by knowledge. Valuations (inside and outside the classroom. Solve worries in class also for students; and teacher socialisation rest of students.	climb to the platform. Participation in forums. Resolution of evaluations at line.	e significant.
Introduction to the calculation differential I	Professor: Design, construction and implementation of the material. Students: access to each of the activities planned.	- Calculate the control of a fiincao. - Set asymptotes (vertical and not vertical). Operate with rational functions. - Define function injective. - Define function inverse. - Characterize inverse function. - Define spinning irrational.	The students visit in platformo cursovirtual where can get to know find the subjectsa develop academically and following the guidelines of the tutor and guides about the course. If take place discussions at	Course visits for development from themes cited. Tasks resolved on digital form and go up to the platform. Participation in forums.	Develops skills inTIC. Gereum Learning Autonomous, collaborative e significant.

Geometry in space	Professor: Design, construgaoe implementation of the material. Students: access to each of the activities planned.	- Determine the domain and contradiction of a Rhinqueo irrational. - Resolve irrational equates. - Determine limit of a function at a point. - Raise indeterminacies. - Calculate the average rate of change of a function on the interval [a,b]. - Calculate, using the definition, the derivative of a function at a point. - Calculate the lateral derivatives at a point of the function. - Apply rules of derivation sum, product, quotient and power. - Calculate the 2^a derivative of a function. - Study the sign variation of the derivative of a function. - Determine the coordinates of the points in the function domain where the 1st derivative changes sign; relative extrema. - Score a point $P(x,°,y°,z°)$ on a system Cartesian, orthogonal monometrical. - Calcularas coordinates of a vector, using equality, (AB) " =B-A - Calculate the coordinates of the	groups of video - classes. Comments on the themes of the course (class). Access to links of other paginasweb (extra class). Knowledge review. Assessments (inside and outside the classroom. Solve concerns in class also for the students; and the teacher socialises the rest of the students. The students visit at platformo virtual course where you can get to know find the subjectsa developing academically and following the tutor's guidance and course guides.	Resolution of the evaluations at line. Course visits for development from themes cited. Tasks solved in digital form and upload to the platform.	Develops skills in ICT. Generate a Learning Autonomous, collaborative and meaningful.

- midpoint of [AB]. - Determine the distance between two points. - Calculate the scalar product of two vectors. - Write equations of the line. - Write plane equations. - Write the Cartesian equation of the median plane of [AB]. - Define analytically spherical surface and sphere. - Resolve systems of three equations with three unknowns by the methods of ordered addition, substitution, mixed and Gauss triangulation. - Study the relative position of lines and planes.	Group discussions of the video lessons are held. Comments on the themes of the course (class). Access to links of other paginasweb (extra class). Knowledge review. Assessments (inside and outside the classroom. They solve class concerns for the students as well; and the teacher socialises the rest of the students.	Participation in forums. Resolution of the online assessments.	

Table 4.3.
Activities related to evaluation criteria.

Activities	Responsible	Objectives	Description	Evidence	Achievements
Evaluation	Teacher: Design, construgaoe material implementation. Students: access to each of the activities raised.	Validate the objectives created at work.	The records provided are tabulated and graphically represented by the platform regarding the frequency of virtual classroom visits, frequencies of accesses by students, frequencies of accesses by activities.	The platform records each and every income made by the students.	To draw conclusions about the impact of the use of ICT on students in grade 10 of the Instituto Politecnico da Graca in Benguela.

4.4.7. Teacher activities.
Table 4.4.
Teacher's activities - assessment criteria.

Thematic Framework	Preparation And Assembly	Resources & Tools	Assessment Tools	Evaluation Criteria
Successes	Construction of Chat, Forums, Wiki, evaluations and other didactic content.	Programmes: Word; Adobe Acrobat; Power Point; Moodle. Meaning: Inserting labels Link to a webpage and archive. Deploy content content package, activities tasks, questionnaires.	Visits to learning content at platform. Tasks solved within the form digital and uploaded to the Platform . Resolution of online and in-class evaluations. Valuations Forums; Chat; Tasks.	Development and fulfilment of the required activities in an agreed time. Participation. Attitude and willingness to work virtually on the established themes.
Introduction to differential calculus I	Construction of Chat, Forums, Wiki, evaluations and other didactic content.	Programmes: Word; Adobe Acrobat; Power Point; Moodle. Meaning: Inserting labels Link to a webpage and archive. Deploy content content package, activities tasks, questionnaires.	Visits to learning content at platform. Tasks solved in digital form and uploaded to the Platform. Resolution of online and in-class assessments Valuations Forums; Chat; Tasks.	Development and fulfilment of the required activities in an agreed time. Participation. Attitude and willingness to work virtually on established themes.
Geometry in space	Construction of Chat, Forums, Wiki, evaluations and other didactic content.	Programmes: Word; Adobe Acrobat; Power Point; Moodle. Meaning: Inserting labels Link to a webpage and archive. Deploy content content package, activities tasks, questionnaires.	Visits to learning content at platform. tasks solved within; digital form and uploaded to the Platform . Resolution of in-line assessments in class Evaluation Forums; Chat; Tasks.	Development and completion of the required activities in agreed Time. Participation. Attitude and willingness to work virtually on established themes.

4.4.8. Description of the teaching-learning methodology.

It is proposed an inductive approach to Mathematics that emphasizes problem-solving methods related to professional media training; based on the development of the following tasks / activities:

Theoretical classes in which the teacher presents and develops the contents of the discipline in an inductive way, based on examples related to the practice of the corresponding professional training. Theoretical results are presented when and as they are necessary, emphasizing how and why they can be applied to a certain problem. The problems are solved, emphasizing the processes involved in the solution methods. The aim is to show how all the competences to be achieved manifest themselves.

Exercises of application of the results presented in classes, which help to understand them. This activity should be carried out frequently and is done in the VLE. Its usefulness is twofold: it allows the teacher to monitor the student and it allows the student to self-evaluate the level of understanding reached. As for its relationship with the competences, it is mainly oriented towards: arguing logically, expressing ideas rigorously and clearly, applying common sense, reasoning quantitatively, handling mathematical language accurately, expressing data graphically, problem solving procedures and solutions, applying a mathematical result correctly, selecting appropriate calculation procedures, selecting appropriate calculation tools and checking that the solution to a problem is correct or at least makes sense.

Through teacher-led Problem Solving Tasks, with the same heuristic approach mentioned above, where the important thing is the processes that lead to the solution. In part of these problems, the use of the VLE is included. It will work as a team. Once again, as in the case of the lecture classes, all the competences pursued come into play.

Informative forums in which real applications of Mathematics are presented in the area of the Environment as the Environmental Management course and in the area of Electricity as the Renewable Energies and Energy and Electrical Installations courses of the Instituto Politecnico da Graga in Benguela, taught by the teacher who teaches the subject of Mathematics. It is oriented to the following skills: translate a real problem into a mathematical problem with data and unknowns, obtain a mathematical model of a real system, design experimental studies useful in solving a problem, interpret physically the solution of a mathematical problem, study and predict the behaviour of a system from the model.

CHAPTER 5
RESULTS AND DISCUSSION.

For the accomplishment of this activity, a documental work was done, in which bibliographic material related to the investigation theme was consulted, as well as articles published in scientific magazines and databases. The didactic concepts of mathematics were emphasized, with focus in giving greater support to mathematics and to the principles of pedagogical action in mathematics; the use of technologies in its pedagogical practice is due, providing the students with the understanding of what they need to know and to learn. Furthermore, the results of the evaluation of the contribution of the virtual learning environment (VLE) as a support to the Mathematics subject for the students of the 11th grade of the Instituto Politecnico da Graga in Benguela (IPGB) are presented. The pedagogical experiment was used as main scientific method and as auxiliary techniques: the focus group technique, and the ladov technique.

For Bogdan and Biklen (1994), detailed data study is the process of examining a set of information in order to reach necessary conclusions that will be useful in achieving objectives. Transcripts of focus group interviews, forums, automated VLE records, questionnaire responses and student-generated artefacts were used to gain a more adequate understanding of these materials and to gain answers to the research problem. A set of analysis techniques was applied, employing systematic strategies through the objectives that report the content of the messages. Bardin (2011).

Being the pre-analysis one of the indispensable steps to reach the proposal, the investigation of the material according to the results process, the deduction and the interpretation conform the stages of the content analysis, this technique is very useful in aspects of the Social Sciences area, taking the human communication as an example. However, it does not imply that there is a systematisation for the different phases. The researcher concentrates continuously on the information collected and the resulting judgements. Pope, Ziebeland and Mays (2006).

The methodological triangulation culminates the chapter, reinforcing the results evaluated in this investigation.

5.1. Evaluation using the focus group technique.

Various authors Aigneren (1998), Gibbs (1997), Verdecia (2011) and Williams and Katz (2001) consider the technical task of focus groups, while Hernandez et al. (2006) determines focus groups in certain sessions, while both coincide in that it is studied through a facilitator who plays the role of moderator where a data collection is redone, in which the differences and knowledge of the same are collected in relation to the variety of approaches.

The focus group technique is carried out by means of an interview script, conducted by a moderator. The aim is for the group to achieve interaction to generate the desired information. A focus group should be composed of 4 to 10 participants, plus a moderator and an observer. The desired information is produced among the participants by listening to different opinions. All this information is processed, considering the data, analysing, understanding and interpreting it, reducing uncertainty, helping to gain a deeper understanding of what is actually happening.

The authors state that each focus group should be composed of 10 members. A comfortable and pleasant environment should be created so that participants can trust each other and express their ideas, experiences, and disagreements freely, allowing the researcher to analyze their attitudes and behaviors.

To form the focus groups, we took into account: the number of members and their performance in the Mathematics subject at the 11th grade of the Instituto Politecnico da Graga in Benguela (IPGB). Groups with a maximum of 10 members were chosen, in addition to reaching two levels of leadership: three focus groups (Annex 7) with 30 people were formed.

For a better execution of these workshops, a thematic guide was used (Annex 8), allowing the coincidences and divergences of the variables of this study to be known. Operational criteria were used for each focal group to analyse the records (Annex 9).

The evaluation of the focus group members of the didactic design and the implementation guide, in the form of coincidences and divergences in the operational criteria, is included in the final report (Annex 10). The following table presents the result of the evaluation of the operational criteria.

Table 5.1.
Final evaluation of the focus group operation criteria.

Theme	Question	Operational criteria
1	1.1	Unanimity of criteria
	1.2	Most criteria
	1.3	Unanimity of criteria
2	2.1	Unanimity of criteria
	2.2	Unanimity of criteria
	2.3	Unanimity of criteria
	2.4	Most criteria
3	3.1	Unanimity of criteria
	3.2	Most criteria
4	4.1	Unanimity of criteria
	4.2	Most criteria

Note. Adapted from Elliot, H. (2005). Guidelines for conducting a Focus group. American Journal For Reserchers, https://assessment.trinity.duke.edu/documents/How_to_Conduct_a_Focus_Group.pdf

With the technique of the focal group it is possible to know that the operational criteria have a great correspondence between the integrants, having that both the virtual learning environment and the guide that was used, both contribute to the integration of the process of teaching mathematics in the students of the eleventh grade of the Instituto Politecnico da Graga in Benguela (IPGB). This allowed the author of this investigation to reflect on the didactic design and the guide.

It is known that with the application of the focal group technique it is revealed how the members agree with the didactic design as well as with the integration in the teaching process of the mathematics subject with the support of the VLE. With these opinions, the author was able to know aspects to reconsider and improve the didactic design beyond the guide, considering the operational criteria not unanimously evaluated.

5.2. Evaluation through a pedagogical experiment.

This pedagogical experiment was designed by Nocedo and Abreu (1984), taking into account the conditions in the teaching and learning processes, in the design of virtual learning environments in support of the mathematics subject for the eleventh grade of the Instituto Politecnico da Graga in Benguela (IPGB).

The pedagogical experiment was carried out from the second semester of the academic year 2019 until the first semester of the year 2020. For the execution of this scientific method, the eleventh class of the Instituto Politecnico da Graga in Benguela (IPGB) was selected.

5.2.1. Characterisation of the teaching group.

For that purpose, a group of students is selected as part of the sample in which the second research task is characterized and it is shown that in this group the interview is applied (Appendix 5.2). In order to know which were the indicators of the dimension of how students use the virtual learning environment through the variable didactic design of the virtual learning environment in support of the subject of mathematics (Annex 3.2).

The group interview produced similar results to those obtained in the characterized sample. The amount of use of the VLE is high, despite showing low satisfaction with the design of the virtual environment. They could not offer criteria for integration with the mathematics subject through the virtual environment, since the course project they developed in the mathematics subject they had just finished was academic.

They agreed with the results of the initial characterization, since a high number of them did not know of the existence of various tools or did not know why to use them. A value higher than 73% affirmed that the use of the VLE was limited to the dissemination and not to the exchange and collaboration. The specialists were part of the sample initially characterized, therefore, it was not necessary to apply any additional instrument to them for inclusion in the experimental group. The above corroborates that the teaching group and the selected experts have similar characteristics to those found in the initial characterization, which is why their selection is considered valid.

5.2.2. Execution results.

For the execution of the pedagogical experiment, several steps were designed in correspondence with the implementation guide. The results obtained are described below.

Stage 1 - Design of the VLE - in the educational context: it was developed during the second semester of the academic year 2019. A work team was formed to guide the execution of the experiment, composed of 4 experts, the author of this research and the two assistant professors of the subject and the psychologist as a panel of experts for content analysis. The experiment was approved by the directors of the Teaching Departments of the discipline, as well as by the management of the Instituto Politecnico da Graga in Benguela

(IPGB).

Stage 2 - Designing the VLE - as an educational context: Next, the dimensions of the VLE as an educational context were designed, resulting in:

- Special anaKtic program for the subject of Mathematics, to be used with the teaching group of the pedagogical experiment.
- Methodological guidelines for the development of the EAP and the use of the VLE for specialists.
- Requirements for the VLE visual design.

Stage 3 - Produpao do AVA: was carried out during the first semester of the 2019 academic year. As a result, the Virtual Course was installed in the Moodle platform of the Instituto Politecnico da Graga in Benguela (IPGB). For the assembly, 32 digital presentations, 11 animations, four e-books (using the aforementioned resources) and a hypermedia report of the most frequently asked questions (FAQ) were designed and produced.

Two textual reports have also been produced in PDF format of case studies of the full cycle of topics, using information from the elementary case studies to be used in the subject.

The content of the main pages of the wikis was designed and produced, on national and foreign websites containing information relevant to the SAP were defined. They were technologically connected.

After these two first stages, the evaluation of the indicators corresponding to the VLE Didactic Design variable in the dimension Teachers participating in the VLE design and the dimension Conceptualization of the VLE Didactic Design showed a significant improvement in all cases, constituting what Nocedo and Abreu (1984) call the second step of a pedagogical experiment.

Stage 4 - Introduction of the VLE - in the SAP of the Mathematics subject: the execution of this stage was carried out during the first semester of the academic year 2019, the second semester of that academic year and the first semester of the academic year 2020. The last two moments corresponded to the teaching of mathematics in the IPGB.

During the months of Margo and April of the first semester of the academic year 2019, the specialists were failed in the use of an AVA in the PEA under a mixed modality. The ADP of the subject was developed with the teaching group of the eleventh grade of the Instituto Politecnico da Graga in Benguela (IPGB), regarding the indicators of the dimensions of the ADP integration variable.

Stage 5 - Evaluation of the VLE: it is considered necessary to clarify that in all the previous stages evaluation axes are developed, which are ways to evaluate the results obtained in each one. However, the latter was conceived to evaluate the results obtained with the pilot introduction, after implementing the didactic design of the virtual environment and executing the didactic strategy of the practical dimension of the VLE as an educational context.

In correspondent with what Nocedo and Abreu (1984) call the sixth stage of the pedagogical experiment, in the stage of evaluation of the VLE, the content analyses were carried out in the months of March and May 2019; before which the experts' supervision was carried out as to the qualitative approach of the scientific research.

The content analysis followed the same design as the characterization, in order to evaluate the final state of the integration of the SAP through the VLE, as a consequence of the implementation of the educational design of the VLE in the Mathematics discipline. These results will be presented below in each of the dimensions and subdimensions of the variable. How to collaborate and interact between teachers and the contents of the virtual learning environment?

Interaction carried out in VLE

We report here the result of the collaboration and interaction during the last year in the VLE in the execution of learning tasks, taking into account its relevance. The aim was to establish the interactions that took place through electronic tools for debates (through messages) and to understand how the collaboration between teachers and contents took place. Then, the interactions teacher-student and student-content will be analyzed, which involve tasks and participation in the construction of the knowledge.

Concerning item 24 of the final questionnaire: do you consider that the communication tools used in the VLE have led to greater teacher-student interaction? This question was elaborated with the purpose of knowing the students' learning about the interaction with the teacher.

The answers to this question are shown in figure 5.1. These results allowed us to know that the teacher-student interaction has been greater with the use of VLE. Where the highest frequency was (76.0%), and the lowest frequency promoted in (3.5%), promoted little (5.5%). Because 92,5% of the students confirm a greater interaction between teachers and students during the use of communication tools in the VLE.

Figura 1.
Teacher-student interaction in the use of communication tools used in VLE.

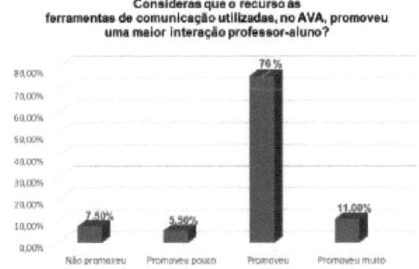

This is the student-content interaction, as presented in the figure:

Figura 2.

Student interaction with the contents of successes and functions.
Figure 5.2 shows information on student-content, successes and fungoes, 698 of the average number of interactions for each student was 6.93, while the average content on successes was 11.63.

Concerning item 25 of the final questionnaire: Taking into account the use of the VLE, do you consider that the interaction with the contents has increased? We intend to know the interaction of the students with the contents. As shown in Figure 5.3.

Figura 3.
Interaction of the student with the contents in the use of the VLE.

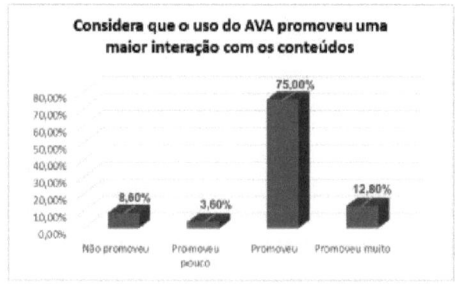

Figure 5.3 shows the results of the question, Do you think that the use of the VLE promoted greater interaction with the content? The highest frequency is 75% who consider that the use of the VLE promoted greater interaction with the contents. Where 12.8% promoted as much interaction with the contents, another 3.6% expressed that they promoted little interaction with the contents, being the lowest frequency. One 8.6 says that it did not promote interaction with the contents.

The focus group is a qualitative research technique. The interview was conducted by a moderator using a script. The interaction between the participants was sought as a method of generating information. The focus group was composed by several students who expressed how the interaction with the contents helped, as they reached a greater understanding and acquired new mathematical concepts previously unknown. For example, the students expressed "the interaction with the contents made it possible for us to work at any time, which made it easier since many times we only had time at night"; another student said "working this way helped me to assimilate contents that many times I could not understand due to the lack of time in the classrooms"; another student said "I had the possibility of working outside class hours, with all the contents, more calmly, it helped me to understand aspects that many times I did not understand during the classroom and now it is easier".

This information gives the possibility to know first hand how the student thinks about his/her interaction with the contents, helps to analyse, understand and interpret this information provided, thus reducing uncertainty, allowing a deeper understanding of the results.

Another subcategory used in the interview was the construction of mathematical knowledge, through discussion and analysis forums in which students participated. For example "having the content of functions helped me to distinguish successes from functions, I realized that they differ only in their respective domains, the domain of successes is the set of natural numbers while the domain of functions is the set of real numbers; having the domain of this content makes it much easier to study it.

In the support classes, it was observed that the students were more interested in the tasks performed in the VLE, showing great motivation during these activities. It was possible to verify the level of commitment of each student, in the work forums where discussions were held between them and the different themes presented by the teacher, being able to appreciate the delivery of the students during the interaction with the contents.

From these results it is understood that the interaction of students with the content contributed to the development and understanding of mathematical knowledge. In this interaction different forms were used: virtual reality, text, sound, video, images. Tools like Power point, Word, pdf, Paint. Interacting with them allowed them to better understand meaningful and transferable learning and the ways of representing knowledge: by deductive, inductive and critical reasoning, by problem solving, by creativity and the possibility of reflecting on their knowledge; and the methods they use when thinking and learning.

Social presence in the VLE

Taking into account the relevance of student-student, student-teacher interaction, they have a social presence in VLEs, because they are considered a fundamental factor for learning specific contents and also social skills, so, if students do not feel isolated, they will be able to learn more, creating environments for social interaction, increasing the feeling of social presence, and making learning easier.

This type of communication is not new for students, what is really new is using mediated learning technologies, using them to learn, discovering how much they can do individually, how much they can help and be helped by other students, making contributions and analysing what other students do, being critical of their own work and that of others, making collective contributions to knowledge.

Online learning communities create the feeling of belonging to a community where learning is strengthened and contributes to the learning of others, creating a dynamic of social interaction of individual and collective growth through the fictive reflection in the investigation of content, a continuous presence of contribution to the continuous knowledge that is built among other individuals, this collaboration is strengthened with an effective, open and coherent communication among the members of this community.

After examining the content of different forums, interaction between teacher, students and content was verified. This interaction favors the individual construction of knowledge, synchronous and asynchronous communication, the possibility of sharing spaces, stimulating a critical perspective, sharing knowledge, investigating, producing in a learning community; this is corroborated by Inacio (2006).

Collaboration

Through the activities and tasks, it was possible to exchange ideas, as well as to develop greater autonomy in solving problems at group level, to develop critical thinking, to contribute ideas, to ask for help, to collaborate, to communicate around mathematical knowledge, which has contributed to the development of students' knowledge, in these subcategories that are integrated in this category.

Different subcategories were developed, the first of which was the participation in the tasks of different individual and collective paths used in the forum through the construction of knowledge generated in asynchronous discussions in the projected learning environment. Example: "... I really didn't think I would make it, it was thanks to you, among all the ones we could"; "... We worked on the graphical representation of successions and functions using the GeoGebra application, I could clearly see that the graph of a succession and the set of isolated points is due to its dominance which differs from the graph of a function ..."; ".... The use of GeoGebra allowed me to understand mathematical relations beyond the statics, besides providing us with numerical, geometrical and algebraic characters".

From the practice of sharing and exchanging ideas, thought processes are developed where students use their own language, a process of construction of the mathematical language, which does not arise by imposition of the teacher. For example, in Task 2, students reported "I must admit that this is the only way I understood, with the help of my classmates and the teacher, since in class everything is very rigid ..."; ".... And what I say, we are a team, we are conquering many things as a group ..."; "It was only when each one of us began to explain task 2 that I understood, we all began to solve the other parts of the question."

It is recognized a participative collaboration where the students could work individually developing self-learning competences, collaborating with other students that could ask questions, see other points of view, besides supporting with their contributions when collaborating to the construction of the knowledge and complement ideas, and of course, having the teacher as a facilitator at all times, helping, explaining, all this to facilitate the teaching-learning processes; Using the VLE as a tool, which facilitated the students to understand mathematics much better, reason for the students to increase the esteem for the subject of mathematics. It allowed the teacher to give a differentiated attention, to measure the progress of each one of them, to help in the understanding of mathematics. These results are the result of a proper use of ICT in education.

The focus group technique reveals how students and their experiences communicate through dialogue, perceptions of their social and community environment, and written reflections. With this technique, attitudes, perceptions, expectations, and behaviors were accessed through exchange. For example, students expressed: "I liked working in groups, you can always ask questions, the work is faster and you also have different opinions and points of view"; "The truth is that this is how you work in a more dynamic way"; "I not only shared ideas, I also learned a lot, thanks to the group work, mathematics is more fun".

This experience allowed the teacher to really know how much she had contributed to the students. If we take into account the students' increased motivation, the ease they find to ask each other questions is not new, but to do it with tools they know but didn't use in the educational environment, it turned out to be a discovery from the beginning. They can develop potential skills with the technologies in their learning environment, giving them a new way where collaboration, communication, interaction play a decisive role in the students' learning and in their new reality.

The use of artefacts produced by the students, in the resolution of the tasks, has highlighted the relationship of the activities with the use of digital artefacts where the students learn by doing in a collaborative way, offering added value to the projects that are developed in the classroom that seeks to enrich the process; with an approach to learning that emerges bringing changes from the knowledge of the student, who learns with their interests and needs.

Considering the learning activities and the artifacts generated by the students, a better resolution of the problems is achieved by showing the development of a critical thinking during the resolution of the problems, not only by solving them, but by recognizing the mistakes made, in the search for understanding the result; an example of this shows a student asking for help for not detecting a mistake in an activity related to the amount of minutes of use mentioned when making a call from a prepaid mobile phone and the amount to pay. In a certain company, if you talk for one minute, you have to pay 70 kzs, if you talk for two minutes, the payment corresponds to 140 kzs and so on. As we can see, student number 2 helps student number 1 by presenting a function relating the variable "amount of minutes spoken" to the variable "amount paid to the company".

1- #1: Hello, I have a problem with task 3, the situation I don't know if it can be represented as a function that relates the variable "number of minutes spoken" with the variable "amount that was paid to the company", can someone help me where I am wrong, please?

I need help from colleagues, does anyone have an answer, it would be of great help to me, thank you in advance. Greetings

Attached file: Task_3.ods

2- # 2: Hello, in this case, the number of minutes spoken will be the independent variable x, and the quantity we cancel will be the dependent variable y = f (x), since it depends on the number of minutes spoken.

Representing this as a function, we have: f (x) = 70x A greeting.

3- #1: Thank you very much, Student #1. It works.

4- - 2: Yes, very good. Remember that f (x) is the amount you paid the company and that 70x is the value of the minute multiplied by the number of minutes spoken.

5- #1: Yes Thank you very much, I get it.

Prof[3] - Remember if you analyze the domain of this function, that is, the set of values that can assume the independent variable assigned by x, you must focus on what this variable represents, in this case the number of minutes. This indicates that x can assume only positive values and zero, therefore, the domain of the function will be the set of non-negative real numbers.

We must analyse the values that the dependent variable f (x) can assume; we must observe that the value f (x) is obtained by multiplying 70 by x, where x will be a positive number, due to this we will only obtain positive values and, therefore, the path of the function will be the set of positive real numbers.

This reflects that the task of student n°2 is modified, assuming the role of teacher by guiding student n°1 in the process of solving the activity. In this sense, it is student n°2 who manages and adjusts. On the other hand, there is a joint contribution of both students, shared by the presentation of a solution method during the collaborative interaction.

Furthermore, the subcategories: group autonomy, sharing and exchange of ideas, critical thinking, communication and construction of mathematical knowledge were analysed. It was possible to observe the students' collaboration in discussion forums.

Here it is necessary to determine in common, which problems are going to be faced immediately and which ones in a mediate way, in order to define the objectives and the tactics to follow, understanding by tactics the steps that are going to be taken, the time, the place and how they are going to give. From here they must continue specifying the possible axes that, as a pedagogical group, they are going to adopt, taking into account the results of the diagnosis and what is proposed by the group. Acosta and Zaragoza (2010 pp. 24-25).

In the interviews with the students, when analysing the collaboration in the group, they revealed how good it was. For example: students "I always rejected the computer, I didn't like it, but it was explained to me, and I understood, I could learn and help others"; "I am really surprised, I never believed I could learn mathematics, thanks to my colleagues"; "I would like this experiment to be carried out in other disciplines". It is recognised that the collaboration went from less to more, as the students acquired skills, the more they became interested.

If we analyse the results of the VLE we can find several tools that can be used for collaborative work in virtual environments, namely: diary, portfolio, interview, forum and the observation technique. The individual contribution of each student to the joint tasks helped to achieve common objectives, such as: social interaction, performance of joint tasks, interdependence and development of communication skills, among others. For the evaluation of the processes, the metacognitive abilities were reached, which were verified through: the revision of the presented tasks, the time that each student used for the development of the activities, allowing the sharing of data and the production of knowledge collectively, amplifying their experiences and stimulating collaboration among their colleagues.

How do you consider the use of VLE to support students' learning of mathematics?

We analyzed the students' results regarding the use of VLE in mathematics learning; in figure 5.4 we analyzed the data from the categories that allowed the interactions performed in VLE, focused on discussions where the development of mathematical knowledge is done around social interaction, so when asked 68.7% responded that they liked it very much, being the most frequent, followed by 16.4% who expressed liking. This means that the total degree of satisfaction equals 85.1%, which is considered positive, as their answers show: Example. Students: "What I like the most is being able to evaluate myself and see my mistakes"; "I was gradually learning more"; "I wish all the subjects were like this, it is really another level"; "I passed the work, but I managed with the help of my colleagues". Other students expressed that they did not like working with the VLE, reflected in 6.2%; 5.6% did not answer the question, the lowest frequency being 3.1% who liked it a little. All this had an average of 14.9%, where students expressed problems with the internet, or with access to the computer.

Figura 4.
The pleasure of using the Moodle platform to carry out group tasks at a distance.

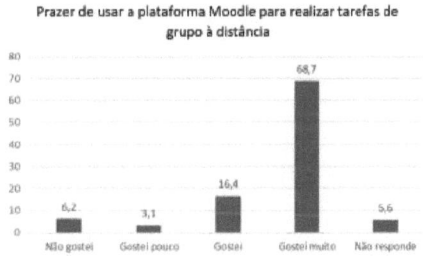

In graph 5.5 it is evident the highest frequencies found in the resolution of tasks, 26.7% of students prefer to have access every day at home; those who do at school equals 26.7%; 20% use in public places often for lack of a computer, the rest is divided by 15% preferring to at home with friends and less often at home with relatives 11.7%.

Figura 5.
Location and frequency of access by students to the VLE.

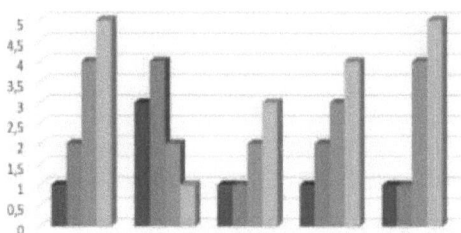

During the execution of the tasks, how often did you access the VLE?

The most common problems encountered when connecting remotely from home were: lack

- Never
- Rarely
- Sometimes
- Every day

of time, 37.4% of the students; 43.3% rarely had difficulties with internet connection; 45.6% never had slow access to the platform; 42.9% always forgot their username and 46.1% almost always forgot their password. platform; 42.9% always forget their username and 46.1% almost always forget their password; according to figure 5.6.

Figura 6.
Regularity of access to the VLE during the performance of remote tasks.

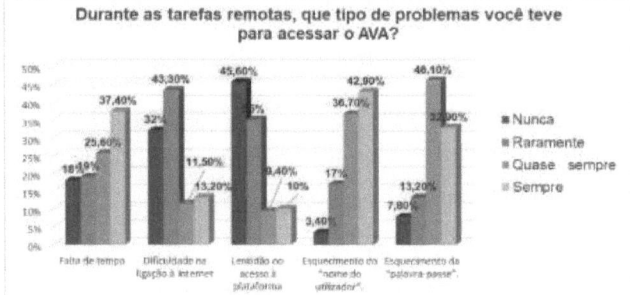

Taking into account the obvious relevance of the use of VLE as a support for teaching in the physical classroom, in order to provide and encourage this process, it was found that 49.1% of the students declared it important, with the highest frequency; 41.2% declared it very important and 4.1%, with the lowest frequency, as not very important.

Figura 7.
Stimulation of students in the teaching process, considering the importance of using VLE to support mathematics.

Importance of the use of VLE como complement the face-to-face teaching, como way to stimulate and favor the teaching process

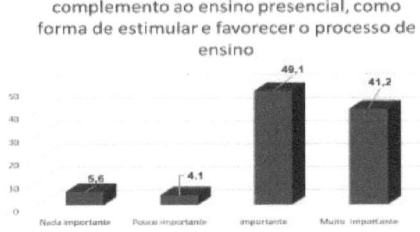

Step 5 - VLE Evaluation: from the beginning this evaluation is carried out, in each developed action the virtual learning environment was evaluated, in order to know if it was in accordance with the achieved results. Already in this phase, the objective is to evaluate the quality of the educational program and the VLE structure, its instructional design, establishing the phases of this process and the criteria to take into account in it and with a pedagogical approach, taking into account the interactions of the teaching - learning process. After implementing the didactic design of the virtual environment and developing the strategies that have contributed to their own learning and ownership of knowledge. The evaluation of learning has been integrated, in addition to evaluating learning. Communication continues to monitor and accompany students, taking into account the feedback they have contributed to the construction of a learning evaluation. In the end, the evaluation gives the possibility to create and implement improvement programs in the VLE, improving the teaching and learning process, which benefits both teachers and students.

For what Nocedo and Abreu (1984) call the sixth stage of the pedagogical experiment, in the phase of evaluation of the VLE, content analyses were carried out in the months of March and May 2019; prior to this, the experts were surpassed in terms of the qualitative approach to scientific research.

The content analysis followed the same design used in the characterization, in order to evaluate the final state of the integration of the EAP through the VLE, as a consequence of the implementation of the educational design of the VLE in Mathematics. These results will be displayed later for each of the dimensions and subdimensions of the variable.

How to collaborate and interact with each other, with teachers and with the content of the virtual learning environment?

Interaction made in VLE

Considering the relevance of collaboration and interaction during a year in the VLE in the realization of learning tasks, we want to establish and understand how the collaboration and interaction between teacher-student and student-student took place, this content is then presented in discussion forums covering the tasks or participation and knowledge construction.

It is verified a total of 58%, in which the students manifested an increase of motivation with the use of the didactic units covered in the VLE, being verified the highest frequency. And the lowest frequency was 9% and that the students stated that they did not feel motivated with the use of the didactic units presented in the VLE.

Figura 8.
The use of the VLE increased their motivation to develop and build their knowledge in the didactic units.
O use of the VLE has increased your motivation to develop and build your knowledge of the didactic units taught

It was noted a sum of 78%, in which students expressed the importance of using the VLE to share information and generate shared knowledge, which was found the highest frequency. The lowest frequency was 4%, where students stated that there was nothing important about using VLE to share information and generate shared knowledge; this information is shown in figure 5.9.

Figura 9.
Importance of using VLE to share information and generate shared knowledge.

According to figure 5.10, there is a frequency of 79.6% in which students recognize the importance of using VLE in the learning process. This was the highest frequency. And the lowest frequency was 6%, where students stated that there is nothing important about using VLE in the learning process.

Figura 10.
Importance of using VLE in the learning process
Importance of using VLE in the learning process

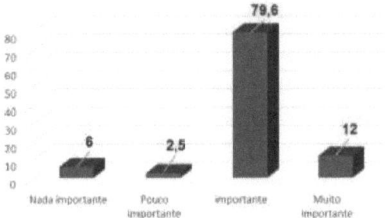

Importância de usar AVA no processo de aprendizagem

For Jonassen (2007), VLEs represent a learning tool, since, by navigating in their environments, interacting, participating, students will be able to collaborate among themselves, building their own knowledge, with the freedom to choose where and how to study. Not only do they want more independence and autonomy in the way they learn, but they also develop investigative skills due to the huge amount of information available to them BECTA (2007).

The VLEs are an interaction tool to support students and provide greater monitoring of student activities to the teacher, stimulate student-teacher communication and strengthen student-student communication, which facilitates the process of teaching and learning, This is supported by Flores and Flores (2007), Lacerda (2007), Lopez and Gomes (2007), Ramos (2005) and others.

In the constructivist perspective, an interaction and exchange of knowledge between teacher and student is proposed with the aim of achieving significant knowledge. The knowledge process is focused on the student. Each person brings with them a knowledge structure in which they assimilate and incorporate other new learning, offering them the possibility to control the direction of their own learning D^az (2012). By using the VLE in the learning process, we sought to propose tasks and activities that present challenges and challenges that can be assumed in their level of knowledge, providing an experience for the individual in the construction of their knowledge.

The students' evaluation of the use of the VLE approved that it facilitated the learning of the contents of the mathematics discipline. The teachers achieved greater control over the activities carried out by the students, presenting an (87%), which contributed to many students overcoming difficulties that were impossible for them until now (71.6%), with the exchange of ideas, it promoted new knowledge (75%), it favoured thinking in the understanding of mathematics in daily life (78,3%), improved their skills in using technologies (84.7%), stimulated their learning of mathematics (82.2%), facilitated their understanding in solving problems and doubts (77%), stimulated a growing interest in mathematics (58.1%), provided remarkable information (70.5%), achieved control of the students' activities (66.2%), promoted responsibility and autonomy in their learning (59.9%), increased student-teacher interaction (85.7%), provided an opportunity to exchange ideas (83.2%), promoted student-teacher interaction (85%), improved student-student interaction (65%), stimulated subject work (72.9%), increased critical thinking (58.4%), promoted mathematical reasoning (75.8%), increased mathematical intercommunication (60.9%), favoured group work, attending opinions and contributions from different actors (65.3%), stimulated the exchange of ideas (85%), increased collaboration between students (81.3%), allowed to highlight the importance of mathematics in daily life (68%), problem solving (85.4%) and encouraged self-learning (81.8%).

Figure 5. 11.
Characterization of the importance of the teacher's role in the VLE.

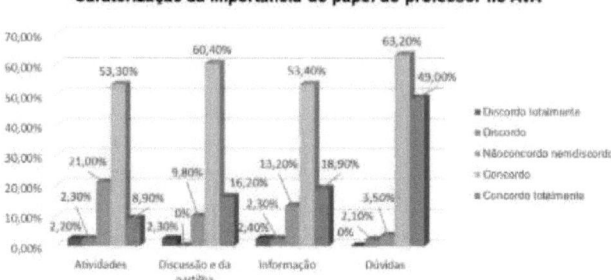

As can be seen in the following figure 5.11, where an analysis was carried out, it shows that 53.3% recognised the role played by the teacher in the learning environment, as she was the one who guaranteed the activities; 60.4% in which there were debates and exchange of ideas about the different topics they worked on; 53.4% reviewed the information presented; 63.2% where it is recognised that the doubts that arose during the work were clarified. The students' opinion was known in relation to the relevance of the role that the teacher plays in the virtual learning environment in the mathematics subject, being a facilitating role where she stimulates, developing critical thinking, Matos (2010, p. 21).

With the use of VLE as a support to the mathematics subject, do you consider that it contributes to mathematics learning?

Based on a pedagogical component, the VLE was built as a collaborative learning environment in an educational context where the student is able to strengthen issues such as reasoning, self-learning, independence and collaborative learning, therefore its motivation with the resolution of cognitive problems propitiated to maintain a positive synergy. It has facilitated and in turn made its use more flexible, as it has allowed the use of mobile phones and computers and to connect from any device and be interconnected in effective time. Favouring the construction of their knowledge through co-authorship, collaboration, greater individual responsibility, and progressively forming critical thinking, enriching and improving their language. To achieve that this environment becomes a space where the resolution of group tasks, favoured with the joint construction of new meanings and that enable group interaction. The resolution of problems assumed in a collaborative form where they socialized and debated, using the platform through the thematic forums, acquired greater relevance in their formation, since they understand the usefulness of those who learn, motivating them to deepen other themes in an autonomous form. This theoretical and conceptual preparation of mathematics is a useful tool for problem solving and decision making. The communication and the interaction succeeded in strengthening the interpersonal relationships between the students, for example, in the case of the interactions, they promoted the creation of learning communities, which allowed the development of cognitive, affective and social processes necessary for this educational process. The social presence, as well as the transactional distance, were strengthened with the communication and interaction that took place in this educational context.

Bruner (1960), in his constructivist perspective, proposed that the student would have more responsibility in the construction of his knowledge, and that the teacher-researcher would be responsible for guiding and enhancing his learning.

The role of the student becomes more active and vital to improving their own learning. Involving them in the decisions of planning their time, location, consultation sources such as databases, blogs, academic networks, among many other possibilities. This encourages self-learning and further study on topics of interest.

1.3. Evaluation by ladov technique.

The ladov technique is named after its creator, V. A. ladov. Researches where this technique has been used, such as Hernandez (2010), Lopez and Gonzalez (2002) and Tejedor (2005), describe it in the study of the level of satisfaction of participants in various training contexts. 60 of the 66 students participating in the experiment and all the project specialists were willing to participate.

The technique consists of three closed questions, interspersed in a questionnaire and whose relationship the respondent does not know. Its aim is to evaluate the level of satisfaction according to what is known as "ladov's logical table". The answer to these three questions makes it possible to place each subject, according to the logical table, on a satisfaction scale and then to calculate the Group Satisfaction Index (GSI) and to evaluate it according to the intervals shown in Figure 5.12.

The use of this technique aimed to evaluate the level of satisfaction of the participants with the VLE after the pedagogical experiment. A questionnaire was applied to students (Annex 11.1) and experts. In both cases, questions three, eight and ten, in that order, were used to compose the logical tables and calculate the GIS.

Figura 12. .
ISG classification intervals [Source : Hernandez (2010)]

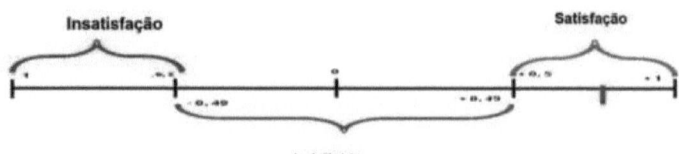

Indefinido

Note. This figure shows a satisfaction scale, the calculation of the Group Satisfaction Index (GSI) and how to evaluate it according to the intervals. Prepared by: HERNANDEZ, S. (2010). Methodische Konzeption Gestaltung von Aufgaben fur studentische Lernen Stadtverwaltung Havanna Agrarian University. https://repositorio.uci.cu/jspui/handle/ident/4588

The distribution of the individual results in the students' group, according to the satisfaction scale, is reflected in Figure 5.13. These results give an ISG of 0.71. This value, between +0.5 and +1, implies satisfaction, since it is interpreted as a positive evaluation of the virtual environment they used and implemented the VLE's educational design.

Figura 13. .
Number of students according to the satisfaction scale.
Students according to the satisfaction scale

The distribution of the individuals' results in the expert group, according to the satisfaction scale, is reflected in figure 5.14.

Figura 14. .
Number of experts according to the satisfaction scale.
Experts according to the scale of satisfaction scale
Especialistas de acordo com a escala de satisfação

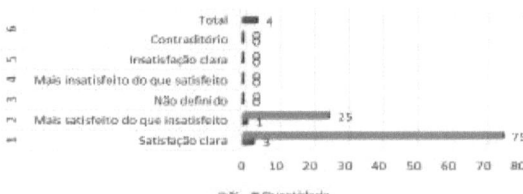

These results produce 0.65. This value, as in the case of students, implies satisfaction, since it is also interpreted as a positive evaluation of the virtual environment. However, the experts in their answers to questions 2 and 1 and 2 issued a set of criteria that contributed to the improvement of the didactic design of the VLE and its implementation guide.

The interviewees were concerned about the pedagogical preparation that teachers need to implement a VLE under the proposed didactic design; as well as the need to graphically represent the didactic strategy that enhances the understanding of it and its use. They also expressed concern about the integration of ICTs into a single technological environment, since several industrial projects deal with classified and confidential information, which cannot be published to members outside them.

All concerns have been resolved. The solution to the first is contained in the teacher enhancement stage of the implementation guide. The second was solved with the representation of the internal structure of the practical dimension of the VLE as an educational context. For the third, the indicator of availability of engineering documentation in the project, without restrictions of use and internal publication in the IPGB, was included as a requirement that an industrial project must meet to be selected.

5.4. Results of the methodological triangulation.

The results obtained in the pedagogical experiment and in the techniques were triangulated to establish their divergences and coincidences. In turn, this methodological action allows us to unify criteria in relation to what Nocedo and Abreu (1984) call the "control experiment" and the last stage of the pedagogical experiment. At the same time, it allows us to evaluate the degree of compliance with the research objective.

Hernandez et al. (2006) classify triangulation into three types: "of the theory or disciplines", "of the researchers" and finally, "of the methods", which responds to the interests of this research and is described as the confrontation of the results obtained by different methods in the search for similarities and differences in the evaluation criteria and thus increase the credibility of the research carried out.

Arias (1999) considers that triangulation occurs within the same method and between methods. Cisterna (2005) and Vera and Villalon (2005) state that the latter type allows complementarity between the methods used, so that the deficiencies of one are diminished by the strengths of the others.

After applying the methodological triangulation, the conclusion is reached that: the transformation of the VLE of the subject of Mathematics, based on its didactic design and on the implementation guide proposed in the research, allows the contribution towards integration to be evaluated as positive between the PEA through the virtual environments of the Polytechnic Institute of Gra?a, in Benguela.

This conclusion is reached because:

1. The indicators of the variable Integration of the EAP in the VLE, obtained in the pedagogical experiment a significant increase in relation to the values of the characterization performed in the second task of the investigation.

2. The focus groups' assessment of the operational criteria in relation to the contribution to integration of educational design and its implementation guide was by no means a minority of criteria.

3. The group satisfaction rates for students and experts were in the satisfactory range.

4. In the content of the compositions made by the students, descriptions, judgements and positive evaluations of the VLE and the EAP are observed.

5.5. Conclusions of the chapter.

☐ The application of the focus group technique allowed a favourable evaluation of the didactic design of the VLE of the subject of Mathematics and its implementation guide, with regard to its contribution to the integration of the EAP, the coherence of the guide in relation to the didactic design that supports it and the systematics of the latter.

☐ The methodological triangulation allowed us to corroborate the results obtained separately by means of the pedagogical experiment and the techniques, with regard to the contribution to the integration of the EAP of the didactic design of the VLE of the Mathematics discipline and its implementation guide.

☐ The pedagogical experiment carried out allowed us to evaluate the transformation that the integration of the PEA suffers in the VLE when its characteristics are modified from the proposed didactic design and the implementation guide is used. The change between the initial and final states of the variable is verified with the increase of the values in each one of its indicators.

Conclusions

This research was carried out with the purpose of using the technologies to support the learning of mathematics through a virtual learning environment as a support to the subject of mathematics for students of the eleventh grade of the Instituto Politecnico da Gra?a in Benguela; a descriptive, exploratory and correlated study, of mixed methodology with theoretical reference where different instruments were applied such as the interview, the observation, the inquiry and techniques such as the focus groups that served to collect first hand data of which participated students as well as teachers and administrators. The results obtained after the application of the instruments will allow us to contribute to new lines of research in virtual environments as a support tool for the training subjects at the Instituto Politecnico da Gra?a in Benguela, as well as at other teaching institutions in Angola.

a) To analyse the main theoretical-methodological references that sustain the support tools for the Mathematics subject of students of the eleventh grade of the Instituto Politecnico da Gra?a in Benguela (IPGB).

In this research, a review of the existing literature was conducted on Virtual Learning Environments as a support for Mathematics, some experiences in America, Africa, Europe and Asia and the challenges of education in Angola, especially in Benguela. The study was based on previous works. Considering the outstanding works on the subject. It is important to mention that the literature in this area is still very scarce in the Angolan context, as well as the existence of scientific publications, which led the author to deepen data collection for this research. The contribution of the Virtual Learning Environment (VLE) as a support to the Mathematics subject for students of the eleventh grade of the Instituto Politecnico da Gra?a in Benguela (IPGB) was analysed.

The theoretical and methodological contributions for the application of resources aimed at improving and extending the students' performance in virtual classes are that many of the problems that affect the quality of teaching in this modality stem from the inadequate transposition of the face-to-face model online and, in many cases, from the institutes that do not consider the real needs for planning these classes.

As the discussion on the subject expands, the challenges of teaching and learning in virtual environments reveal countless reasons to contemplate the use of computer applications for learning as an essential object in an attempt to create conducive so that students feel motivated and challenged to dive into the proposed content, providing elements that significantly recognize the importance of study and deepening in the theme proposed by the teacher. Given the above, what was sought in this research was to deepen the studies on theoretical and methodological considerations for the production of teaching materials in virtual learning environments and, consequently, to analyze computational applications, as well as a form of didactic construction, although seen as an element of contextualization and / or auxiliary order. It was found evidence that computer resources as elements of contextualization and / or thematic synthesizers create conditions for the students of the eleventh grade of the Instituto Politecnico da Graga in Benguela (IPGB) to make spontaneous decisions to deepen the theoretical material of the curricular unit or even explore the complementary materials indicated by the teacher.

b) Review bibliographic material from the Ministry of National Education, (curriculum guidelines, competency standards), which will support the development of the project.

Education in Angola has gone through times marked by advances and setbacks in educational policy that have conditioned its positive evolution, through bibliographic research and consultation of archives, highlighting the most emblematic measures. During the development of this dissertation, it was initially analyzed the existing literature on support methods to the subject of Mathematics, of the students of the eleventh class of the Polytechnic Institute of Graga in Benguela (IPGB) and bibliographic materials of the Ministry of National Education (curricular guidelines, (curriculum guidelines, competence standards), which allowed supporting and developing the project, such as changes in the educational processes, ICT and teaching-learning methods, evolution of VLE and current trends and their didactic development, mathematical competences developed in VLE.

c) To characterize the didactic design of the VLE as a support tool for the subject of Mathematics for students of the eleventh grade of the Instituto Politecnico da Graga in Benguela (IPGB).

The didactic design was proposed with a constructivist approach, having the teacher as a guide and motivator of the learning process, applying the strategy of analysis of technological products in the development of the theme of mechanical operators. The selected mediating tool was the Moodle platform, which allows the development of socio-constructivist activities such as: Forums, Chat, Interactive activities and the interaction of the student with his colleagues and the teacher, who in this case presents himself as a mediator of knowledge and use of VLE tools. The validation of the proposed learning environment was worked with a quasi-experiment, applying the environment mediated by VLE to the students of the experimental group, while the traditional methodology was used in the control group. Then, the validation was performed through a test that the group presented and the results obtained were analyzed to test the functionality of the environment. The analysis of the results showed a better performance in the students of the experimental group in relation to the students of the control group. That is, a significant difference was obtained in the academic performance of the group to which the environment mediated by VLE was applied. This allowed us to conclude that the use of the methodology mediated through the VLE had a positive influence in the learning of mathematics.

[a] With the results obtained with this experiment, it was possible to demonstrate that the use of Virtual Learning Environments contributes positively to the process of teaching and learning as a tool to support the subject of Mathematics for students of the 11th grade at the Instituto Politecnico da Graga in Benguela (IPGB); Since, on the one hand, the students

managed to develop the proposed competences, the selected mediating tool was applied, the Moodle platform, which allowed them to develop collaborative and active learning, which captured their attention, motivating them to construct their own learning, managing to overcome the difficulties presented to them during the work. In the same way, for the teacher who participated in the research, it meant an opportunity to renew his pedagogical practices, going from the simple mathematics chair to the use of ICT as a means of teaching that allows classes with greater participation through the opening of spaces that alternate communication with their students, making the teaching process easier and more fun.

d) To evaluate and appropriate different resources and technological tools to improve the teaching-learning processes and develop students' random mathematical thinking.

The application of the focal group technique, the experts' criteria and the comparison of their results through methodological triangulation, allowed corroborating the quality, effectiveness and validity of the didactic design of the VLE in terms of offering instructional and educational potential that can be obtained from the use of ICT in the teaching and learning process of the subject.

Currently, it cannot be said that there is complete unity and cohesion in terms of evaluation methodologies. Different traditions and assessment cultures coexist. Due to the different origin of the teachers' disciplinary field, and due to the heterogeneous nature of the contents. The framework proposed in the programme makes this state of affairs possible, so that if different assessment methodologies are not adopted in an open wide range of traditional ones, on the one hand they are still effective, thanks to the know-how implicit in the analysis of texts, practices, etc.

The focal group technique provided all the necessary information for the production of the virtual learning environment in support of the subject of mathematics, demonstrating the same pedagogical development in the virtual environment achieved the implementation and integration of the teaching and learning processes in the subject of mathematics where the students of the eleventh grade of the Instituto Politecnico da Graga in Benguela, achieved a better understanding in the learning of this subject.

[a] Given the characteristics of learning evaluation in virtual environments, where different resources can be developed to improve the teaching-learning processes to develop the random mathematical thinking of the students of the 11th grade at the Instituto Politecnico da Graga in Benguela (IPGB), allowing a variety of evaluation mechanisms according to the didactic design selected. The structure allowed to attend primarily, the interactions between teacher-students, students-students, using the different resources that facilitated them: a didactic dialogue mediated by materials, experiences, feedbacks.

With the purpose of obtaining information, the focus group technique was applied to the students, having the opportunity to express their opinion and to express as individuals their way of thinking and feeling about the interaction with the VLE, which caused self-explanations, facilitating the obtaining of qualitative data. This technique, carried out in the form of a group interview, facilitated communication between the researcher and the students. Among their opinions, the affirmative ones prevailed, besides showing motivation for the way they managed to construct their own knowledge, in many cases making comparisons with face-to-face, expressing how they managed to understand in the VLE what they did not achieve in face-to-face, since the interaction between their colleagues through forums and the artefacts produced by themselves facilitated a much better understanding of others than acquiring mathematical concepts that they did not pay attention to before. During the interview, students also expressed themselves in relation to collaborative work, where they mentioned how easy it was to achieve a common goal, where they shared ideas and skills, where two heads think more than one, i.e., the accomplishment of tasks and problem solving, encouraged a culture of learning to develop skills to increase mathematical content. Collaborative work provided the student with autonomy of critical thinking during their performance, allowing greater understanding of the contents, favouring significant learning experiences.

In the evaluation by the Iadov technique, the respondents were concerned about the pedagogical preparation that teachers need to implement a VLE under the proposed didactic design; these concerns were solved. The first solution is contained in the didactic improvement step of the implementation guide. The second was solved with the representation of the internal structure of the practical dimension of the VLE as an educational context. For the third one, the indicator of availability of engineering documentation in the project, without restrictions of use and internal publication in the IPGB, was included as a requirement for the selection of an industrial project. The transformation of the VLE of the Mathematics discipline, based on its didactic design and on its proposed implementation guide in the investigation, allowed us to evaluate as positive the contribution to the integration between the teaching and learning process through the VLE of the Instituto Politecnico da Graga, in Benguela.

Existence of a work plan for the mathematics subject with a common format, containing the guidelines and assessment criteria. Collaborative assessment based on collective or group production, as occurs on wikis, collaborative work platforms or in a broad sense in group work. In this case, the work of all was guaranteed, either by the individual production from the collective work, or by the participation of the evaluator in the process. The interactive evaluation, which is based on the production of each student that occurs in a complex interaction environment, such as a discussion list, a forum or any other asynchronous communication support. For her, besides evaluating the quality of the productions, the interventions should be taken into account from parameters such as relevance and pertinence. Taking into account several factors, including feedback in the evaluation communication, new factors and dimensions are proposed to evaluate.

e) To elaborate the didactic design of a VLE as a support tool for the subject of Mathematics for students of the eleventh grade of the Instituto Politecnico da Graga in

Benguela (IPGB).

Elaboration of the didactic project of a VLE as a support tool for the Mathematics discipline of the eleventh grade students of the Instituto Politecnico da Graga in Benguela (IPGB); to say that 27 classes took place on the virtual Moodle platform, which integrates the course Mathematics on successes; Introdution to differential calculus I; Geometry in space, with the respective learning guides, documents, educational videos, development of lectures in the virtual classroom to debate the themes of each unit, access to the virtual forum and respective evaluations; taking into account the area of Environment, with the course of Environmental Management and the area of Electricity with the courses of Renewable Energies and Energy and Electrical Installations, being evaluated through the platform of the same school on the same subject, but with the support of a virtual platform. The analysis of the results carried out was of vital importance for the project, making it possible to certify the benefits of an education based on the progress of new technologies with the virtual platform in the resolution of problematic situations and simultaneously in the systematization of the contents so that it is impactful in the real practice of the students' daily life.

With the objective of contributing to the teaching-learning of mathematics for students of the eleventh grade of the Instituto Politecnico da Graga, in Benguela. The software selected for the virtual course was moodle, because it is free and useful. Content design: 3 concepts were raised in the form of didactic statistics, which were selected according to the programming in the area of mathematics according to the standards of basic skills issued by the national Ministry of Education. The target-population (students of the eleventh grade) in relation to the management of the virtual moodle platform.

The development of Chat in a virtual classroom for the discussion of the topics of each unit. Access to the virtual Forum for deliberation in digital format and respective uploading of answers to the platform. Virtual presentation of videos for further discussion in class and submission in the notebook of the respective summaries on the theme of the videos watched. Application and deepening of concepts developed in class through videos and simulations on the virtual platform. Wiki revision and support.

f) To verify the practical contribution of the didactic design of a VLE as a support tool for the Mathematics subject of the eleventh grade students of the Instituto Politecnico da Graga in Benguela (IPGB).

The didactic design of the virtual environment presented in the research power systematically instructive and educational for having technologies and services that facilitate self-training, cognitive independence, exchange, socialization, collaboration, collective construction of knowledge and feedback in the communication that is established among the participants. The requirements that guide the didactic design of the VLE are fundamentally related to the availability, possibility of immediate feedback, offering mechanisms for students to participate actively in the construction of learning and, in the permissibility of communication, interaction, interactivity and collective work for participants.

The theory that supports the study carried out, highlights the value of the interactions since they are present in the various activities of the students in training and these were carried out with technological means that require the use of indirect language. The interactions that occurred in this case allowed the formation and reaffirmation of meanings, which reinforced the individual and collective points of view associated with the theme that was the centre of these interactions, which contributed to the accomplishment of the didactic formation of the participants in the field of mathematics education.

Within the ethical component, common and different points of view were expressed around the theme of the discussion and in the interactions a learning process was produced and/or reaffirmed in which mutual respect prevailed, the exchange of academic points of view and motivation, learning and adaptation through the exchange of rules and beliefs prevailed until a common position (consensus) was reached on the proposal that centred the discussion. This can be interpreted as a contribution that can be made to the professional training of mathematics teachers through the use of technological resources.

Once the VLE was implemented to support the subject of mathematics for the students of the 11th grade of the IPGB, the positive result was the promotion of direct interactions, which occur when individuals with the same interests or indirect interests interact - when objects or instruments where signs mediate, such as the use of an interactive platform as the one used in this experiment - allow them to relate better with their environment, with the school contents, and through them their knowledge system is deepened and a more solid and consolidated learning is achieved.

In this process of formation of the students of the IPGB 11th grade, the interactions play a fundamental role, which coincides with several investigations that have been developed in this respect in the last years. From an ethical point of view, the interpersonal relationships which were established in the development of the various interactions which took place, based on the characteristics of the members of the group and their interests in the didactic aspect which was used as the centre of the interactions, to a certain extent, highlighted the culture of the group and its transmission through the process of socialisation which took place.

The first results collected in the work, from the application of the questionnaire to the development of the experiment itself, indicate the scarce knowledge in the management of technologies, namely in the use of forums, Chat in virtual contexts, the first as a means of communication and interaction and the second for the development of learning activities in mathematics topics.

The use of the forum allowed the discussion of ideas and agreement of points of view, one of the central objectives of the research, is centred on the participation and interaction between mathematics teachers, when discussing a set of learning activities about a mathematics unit, particularly the subject of successes and functions.

It was possible to isolate the participations and interactions, making a characterization and analysis of these interactions, from two points of view, by time periods and by individual participation. In the analysis by time periods, it was possible to appreciate the participations and interactions addressed to their fellow teachers, the contents addressed and to whom it was addressed. In the case of the individual analysis, it was possible to evaluate the number of participations of each teacher, to whom it was directed and on which themes the participation was.

The participations and interactions in a virtual forum fomented the discussion and contributed in agreement with the theoretical orientation for the development of abilities and professional competences of the students of the mathematics discipline, using the tools and improving the communication and interacting in a prepared virtual environment. Thus, the students had the opportunity to analyse their own ideas and those of their colleagues and to rethink new proposals, related to the programmatic content, by knowing and manipulating dynamic tools such as the chat and a virtual medium known as forum; The forum was used as a means of communication, which allowed to initiate a discussion around a set of tasks, which allowed students to select a theme of their choice for the realization of these tasks, their participation and interaction with colleagues, so that their professional knowledge was influenced by the knowledge of their peers, who contributed to a socialized knowledge supported by the forum as a means, recording their participation and responses in writing.

The Forum can be a strong resource, if teachers are convinced to use it as a tool to serve professional skills, since it allows them to communicate with other colleagues, to carry out the tasks and activities that other colleagues do in remote locations. In other words, it can be assumed that in future work and its implementation both in teacher training and in the learning of mathematics students, there is much or all to be done, which tells us that many investigations will be carried out in the immediate future. Work along these lines and in various applications will have to result from both research work and successful practice in the same classrooms.

By analyzing the data, it can be seen how the students agree with the benefits that are provided to them in the construction of their learning of mathematics, strengthening the communication and a stronger interaction with the teacher and students, creating the possibility to debate, question, clarify their doubts, helping to develop the facilitator work of the teacher, using a favorable environment known by them, in the interaction through forums, chats, where they addressed issues related in the activities planned through the tasks. To contribute and share new knowledge, where the management of individual and collaborative learning develops reasoning skills, stimulating critical thinking, through the resolution of mathematical problems, allowing the interaction student-ICT being protagonist of its own learning. Helping to a better understanding of mathematics, motivating a greater interest for the subject, reaching a new vision of mathematics in everyday life.

The fundamental limitation for the development of the experiment was the time employed in its development and the frequency of the participants, which was reduced, concentrating mainly on a part of the participating students. The teachers with the greatest number of participants coincide with those who have greater mastery of the use of technological means, an aspect that at the beginning of the experiment did not appear to be widespread and which motivated the inclusion of initial familiarization activities with this type of work.

From a prospective point of view, it is considered that it is necessary to continue searching for ways to carry out research activities with the use of technology, to guarantee a more solid empirical consistency of some results visualised in this work, as well as the verification of the hypotheses that were derived from it.

Recommendations

1. To suggest, based on the results of the pedagogical experience, to the Central Department of Mathematics Teaching of the IPGB, to generalize the didactic development of the VLE for all the disciplines generating future researches in the area of Capacita^ao, validating its use in other technical disciplines of the curriculum and in the improvement of teachers and specialists of the institution.

2. It is recommended to carry out this type of research in other teaching institutes and universities to approach the problem from different contexts and perspectives in order to be able to guarantee better and efficient induction processes and, therefore, better teaching-learning processes, improving educational conduct and desertion rates.

BIBLIOGRAPHY

Acosta, R.C. and Zaragoza, A. (2010). La estrategia pedagogica para el desarrollo de los grupos estudiantiles. Rev Inform Educ Med Audiovis. 7(14): 22-28. https://www.fi.uba.ar/laboratorios/lie/Revista/Articulos/070714/A3mar2010.pdf

ADAMS, N. and DE VANEY, T. (2009). Measuring conditions conductive to knowledge development in Virtual Learning Environments: initial development of a model - based survey. Journal of Technology, Learning and Assessment, 8 (1). [Consulted on: 3 September 2021].
https://ejournals.bc.edu/index.php/jtla/article/view/1620/1464,

Addine, F. (2004). ¡Didactica!^ Que didactica. Didactica: teoria y practica. La Habana: Editorial Pueblo y Educacion, (pp.1-5). Disponivel em: https://scholar.google.com/scholar?cluster=11307544310978416889&hl=en&oi=sch olarr#d=gs cit&u=%2Fscholar%3Fq%3Dinfo%3A-ZDbEaR37JwJ%3Ascholar.google.com%2F%26output%3Dcite%26scirp%3D0%26s cfhb%3D1%26hl%3Den

ADDINE, F. and GARCIA, G. (2009). Componentes del proceso de ensenanza - aprendizaje. In COLECTIVO DE AUTORES, Temas de introduccion a la formacion pedagogica. La Habana: Pueblo y Educacion, (pp. 158 - 170). [Consulted on: 3 September2021].
https://books.google.es/books?hl=es&lr=&id=j9UREAAAQBAJ&oi=fnd&pg=PA157& ots=F86JSN8Ii&sig=Y5mfX1jmw8EB7bgFayFyJ0b6IB0#v=onepage&q&f=false

Adell, J. & Castaneda, L. (2012). Emerging technologies, emerging pedagogies? J. Hernandez, M. Pennesi, D. Sobrino & A. Vazquez. Emerging trends in ICT education. Barcelona, Spiral Association, Education and Technology, (pp. 13-32). [Consulted on: 3 September 2021].
https://digitum.um.es/digitum/bitstream/10201/29916/VAdell Castaneda emergent es2012.pdf

AIGNEREN, M. (1998). La tecnica de recoleccion de informacion mediante los grupos focales [en linea]. [Consulted on: 3 September 2021].
https://es.scribd.com/document/259965263/LA-TECNICA-DE-RECOLECCION-DE-INFORMACION-MEDIANTE-LOS-GRUPOS-FOCALES

Ainsworth, S. A. (2006). Marco conceptual, consider el aprendizaje multiples representaciones, Aprendizaje e Instruccion: (16), 183-198. [Consulted: 3 September 2021].
https://doi.org/10.1016/j.learninstruc.2006.03.001

ALFONSO, I., GARCA, A. and LAURENCIO, A. (2006). Una propuesta alternativa para el desarrollo de la educacion virtual en países en v^a de desarrollo. In COLECTIVO DE AUTORES, Avances y perspectivas de la investigacion universitaria. La Habana: Felix Varela, (pp. 96 - 117). [Consulted on: 3 September 2021].

Almeida, L. (2010). Web 2.0 and learning math standards. (Dissertation Master's Degree), University Aveiro, Aveiro. [Consulted on: 3 September 2021].

https://core.ac.uk/download/pdf/15569234.pdf
Almeida, L., & Freire, T. (2008). Research methodology Psychology Education (5 ed.). Braga: Psiquilfbrios Editions [Consulted on: 3 September 2021]. https://www.researchgate.net/publication/263654768 Almeida LS Freire T 2008 Metodologia da InvestigacaoemPsicologiaeEducacaoBragaPsiquilibrios5 Edicao
ALMEIDA, M. E. B. (2003). The Distance education Internet: approaches and contributions digital learning environments. Educ. Pesqui. 29(2), 327-340. [Accessed: 23 September 2021]. https://www.redalyc.org/pdf/298/29829210.pdf
ALVAREZ, C. (1996). La escuela en la vida. La Habana: Felix Varela. [Accessed: 23 September 2021]. http://www.conectadel.org/wp-content/uploads/downloads/2013/03/Laescuela en la vida C Alvarez.pdf
ALVAREZ, C. (1999). La pedagog^a como ciencia. Santiago de Cuba: Centro de Estudios de Education Superior Manuel F. Gran. [Accessed: 23 September 2021]. https://issuu.com/erickdiaz854/docs/la pedagogia como ciencia. alvarez
ALVAREZ, E., RODRfGUEZ, A. and RIBEIRO, F. (2011). Blended training ecosystems - learning emprender collaborate universidad. Valuing students about resources (Vol. 12 n° 4). [Accessed: 23 September 2021] https://dialnet.unirioja.es/servlet/articulo?codigo=3798494
ALVAREZ, E. and RODR^GUEZ, A. (2011). Learning to Empreender universidad Ecosistemas Formation blanded - learning. Orlando, Florida, USA: SIECI [en linea]. [Accessed on: 23de Septemberde2021] https://www.iiis.org/CDs2011/CD2011CSC/SIECI 2011/PapersPdf/XA548XY.pdf
ALVAREZ, E., RODRfGUEZ, A. and PEREZ - BUSTAMANTE, O. (2012). Ecosistemas de formation transdisciplinar en la universidad: Diseno, implementation y evaluation. Orlando, Florida, USA: SIECI. [Accessed: 23 September 2021] https://www.iiis.org/CDs2012/CD2012SCI/SIECI 2012/PapersPdf/XA912ZX.pdf
ANAYA, K. (2004). A teaching model - virtual learning: analysis, design and application of the Mexican university system. UG. [Accessed: 23 September 2021] https://dialnet.unirioja.es/servlet/tesis?codigo=19507
Arnal, B. (1992). "Adopta la idea de que la teoria crftica es una ciencia social que no es puramente empirica. [Accessed: 23 September 2021] https://orcid.org/0000- 0003-4421-1144
AREA, M. (1998). Los medios de ensenanza: conceptualizacion y tipologia [Accessed on: 23 September 2021] https://ced.enallt.unam.mx/blogs/socio- pragmatica/files/2013/06/Manuel-Moreira1.pdf
Arias, C.E. (1999). Utilizacion del heno de Clitoria ternatea L. en la alimentacion de terneros de leche de doble proposito. CN. [Accessed: 23 September 2021] https://agris.fao.org/agris-search/search.do?recordID=DJ2012037088
Arias Cabezas, J.M., Arranz San Jose, J.M. and Lobo Paradineiro, M.C. (2008). Training

Investigation on Information Technologies Communication Mathematics for ESO Bachillerato. Proyecto Matematicas- TIC Castilla y Leon 06/08. [Accessed on: 23 September 2021] https://infoymate.es/investiga/publicacion.pdf

ATHANASIOS, T. (2007). Description of a Virtual Learning Environment for preliminary schools. VC. [Accessed: 23 September 2021] https://doi.org/10.1177/0047239520908838

Ausebel, D. (1998). Psicológ^a educativa. Un punto de vista cognoscitivo. Mexico: Trillas. [Accessed on: 23 September 2021] https://www.iberlibro.com/Psicolog%C3%ADa-educativa-punto-vista-cognoscitivo-David/22453667993/bd

Bardin, L. (2011). Analise de Conteudo. Lisboa: Edi^oes 70. [Accessed on: 23 September 2021] http://www.reveduc.ufscar.br/index.php/reveduc/article/view/291

Barros, J. A. (2008). Teaching sciences from a didactic perspective French school. Revista EIA, 10, 55-71. [Accessed: 23 September 2021] https://www.researchgate.net/publication/28293293_Ensenanza_de_las_ciencias_desde_unamiradaladidacticadela escuelafrancesa

Bartholomew, K. (1990). Avoidance of Intimacy: An Attachment Perspective. Journal of Social and Personal Relationships, 7(2), 147178. [Accessed: 23 September 2021] https://doi.org/10.1177/0265407590072001

Bautista, C. N. P. (2011). Qualitative investigation process. Epistemology, methodology applications. Editorial Manual Moderno. Bogota. [Accessed: 23 September 2021] https://www.pedagogicomadrededios.edu.pe/wp-content/uploads/2020/10/Proceso-de-la-investigacion-cualitativa.pdf

BECTA, (2007). The impact of ICT in schools: A landscape review. Coventry: BECTA. [Accessed on: 23 September 2021] https://oei.org.ar/ibertic/evaluacion/sites/default/files/biblioteca/33_impact_ict_in_scho_ols.pdf

Benedito, E. (2000). Didactica de la matematica moderna. Mexico: Trillas. [Accessed: 23 September 2021] https://doi.org/10.1080/00207543.2012.655863

Benitez, A. A. (2011). It is important for contextualized events to develop mathematical competences. Latin American Educational Mathematics Committee. C.; (pp. 51- 59). [Accessed on: 23 September 2021] https://core.ac.uk/download/pdf/33251338.pdf

BLANCO, A. (2003). Hipotesis, variables y dimensiones en la investigacion educativa. In COLECTIVO DE AUTORES, Educational research methodology. Current controversial challenges. La Habana: Felix Varela, (pp. 168 - 177). [Accessed on: 23 September 2021]. https://dokumen.pub/metodologia-de-la-investigacion-educacional-desafios-y-polemicas-actuales-9592584192.html

Bogdan, R. and Biklen, S. (1994). Qualitative Research Education Introduction to Theory and Methods. Porto: Porto Editora. [Accessed: 23 September 2021]

https://ria.ufrn.br/123456789/1119
Bruner, J. (1960). "La disponibilidad para aprender", en Palacios, (1988): Desarrollo Cognitivo Press, Cambridge, Massachusetts, EEUU. (Traducida al Castellano por Uteha, 1969, Mexico). [Accessed on: 23 September 2021] https://es.scribd.com/document/454812262/Cap-VII-la-disponibilidad-para- learn-Bruner
BRYMAN, A. (2006). Integrating quantitative and qualitative research: how is it done? Qualitative Research, 1(6), 97-113. [Accessed: 23 September 2021] https://journals.sagepub.com/doi/10.1177/1468794106058877
BULLEN, M. A. (1997). Estudio de caso participation pensamiento crítico nivel universitario entrega conferencias por computadora. UCB. [Accessed: 23 September 2021] https://open.library.ubc.ca/soa/clRcle/collections/ubctheses/831/items/1.0056000
CABERO, J. (1996). Nuevas tecnolog^as, comunicacion y education. [Accessed on: 23 September 2021] DOI: 10.21556 / edutec.1996.1.576
CALDWELL, B.J. and SPINKS, J.M. (1986). Ecole de planification de l'elaboration des politiquesEfficacite. Departement de l'education Tasmanie, Hobart. [Accessed: 23 September 2021] https://files.eric.ed.gov/fulltext/ED352713.pdf
CALLAGHAN, M., McCUSKER, K., LOSADA, J., HARKIN, J. and WILSON, S. (2009). Engineering Education Island: Teaching engineering in virtual worlds. ITALICS [en lfnea] (Vol. 8, n°. 3). [Accessed: 23 September 2021] https://www.researchgate.net/publication/228910166   EngineeringEducationIsla ndTeachingEngineeringinVirtualWorlds
CAMPBELL, D. T. and FISKE, D. W. (1959). Validation convergente et discriminante multitraitmatrice multimethode. Bulletin psychologique, 56, 81-105. [Accessed: 23 September 2021] https://psycnet.apa.org/record/1960-00103-001
CAMPISTROUS, L. and RIZO, C. (2003). Indicateurs de rechercheen education. COLLECTIF D'AUTEURS, Methodologie d'investigation pedagogique. Defis controverses actuels. La Habana: Felix Varela, (pp. 138 - 167). [Accessed: 23 September 2021] https://dokumen.pub/metodologia-de-la-investigacion-educacional-desafios-y-   polemicas-actuales-9592584192.html
Carrilho, C. (2006). The WWW Learning Math Scope StudyAccompanied. UA. [Accessed on: 23 September 2021]. https://ria.ua.Pt/bitstream/10773/9105/1/%282007%29%20A%20WWW%20na%20 aprendizagem%20da%20matem%C3%A1tica%20no%20%C3%A2mbito%20do%2 0estudo%20acompanhado.pdf
CARTELLI, A., STANSFIELD, M., CONNOLLY, T., JIMOYIANNIS, A., MAGALHAES, H. and MAILLET, K. (2008). Hacia el desarrollo de un nuevo modelo de mejores practicas de construccion de conocimiento de campus virtuales. Educacion en tecnolog^a de la informacion de revistas, 7, 45-68. [Accessed: 23 September 2021] http://www.jite.org/documents/Vol7/JITEv7p121-134Cartelli397.pdf

Castaneda, A. E. & Fernandez de Alaiza, V. (2002). Applications Information Technology Communications in the teaching-learning process: Universidad Tecnica de Ambato. [Accessed: 23 September 2021] DOI: 0000-0002-6990-5428

CASTELLANOS, D., CASTELLANOS, B., LLIVINA, M. and MORENO, M. (2009). Aprendizaje y desarrollo. In COLECTIVO DE AUTORES, Temas de introduccion a la formacion pedagogica. La Habana: Pueblo y Educacion, (pp. 291 - 315). [Accessed on: 23 September2021]. https://books.google.co.ao/books/about/Temas_de_introducci%C3%B3n_a_la_formaci%C3%B3np.html?id=j9UREAAQBAJ&redir_esc=y#:~:text=La%20psicolog%C3%ADa%20pedag%C3%B3gica%2C%20la%20social,al%20servicio%20de%20su%20perfeccionamiento

Castro, F. R. (2004). New challenges initial teacher training from a curricular perspective. Horizontes Educacionales. Chile: Issue 9, 17-23. [Accessed: 23 September 2021] https://dialnet.unirioja.es/servlet/articulo?codigo=3993324

CEMILE, F. (2008). A multi - agent adaptive learning system for distance education. METU [Accessed: 23 September 2021] https://www.semanticscholar.org/paper/A-MULTI-AGENT-ADAPTIVE-LEARNING-SYSTEM-FOR-DISTANCE-Cem/bef6e4c4541d2aca91d0de0f1c9cff09def02eedc

CHARD, S. (2011). Construire une classe virtuelle : environnement educatif pour la generation Internet. CU. [Accessed: 23 September 2021] https://www.semanticscholar.org/paper/Building-a-virtual-classroom-%3A-an-education-for-the-Chard/5df257de4226c572336f0192d5f973b3cb0cf81c

CISTERNA, F. (2005). Categorizacion y triangulacion como procesos de validacion del conocimiento en investigacion cualitativa. Teona [en lfnea] (Vol. 14, n°. 1). [Accessed on: 23 September 2021] https://www.researchgate.net/publication/26422891_Categorizacionytriangulacion_como_procesos_de_validaciondel_conocimientoeninvestigacioncualitativa

Cocunubo-Suarez, J. I., Parra - Valencia, J. A. and Otalora-Luna, J. E. (2018). Evaluation proposal Virtual Environments Teaching Learning based on standards Usability, Technological, 21 (41), 135-147. Technological ISSN-p 0123-7799, [Consulted on: 1de September2021]. https://www.redalyc.org/journal/3442/344255038008/344255038008.pdf

COLLAZO, R. (2004). Una concepcion teorico - metodologica para la production de cursos a distancia basados en el uso de las tecnolog^as de la información y las comunicaciones. ISP. [Accessed 23 September 2021] https://www.researchgate.net/publication/236574917_Referentesteoricosdelae_ducacionadistancia

COLLAZO, R. (2009). Los medios en el proceso de ensenanza - aprendizaje. En E. HERRERA, & R. COLLAZO, (comp.), Preparacion pedagogica para profesores de la Nueva Universidad Cubana, 2ª edition. La Habana: Felix Varela, (pp. 73 - 84). [Accessed on: 23

September 2021] http://www.biblioteca.uh.cu/nuevas- adquisiciones/preparacion-pedagogica-profesores-nueva-universidad-cubana

Cordoba, A. (2015). ICT-mediated learning environments proposed monitoring of the Colegio Universidad Pontificia Bolivariana Mathematics area. Medellin: Universidad Pontificia Bolivariana. [Accessed: 23 September 2021] https://repository.upb.edu.co/bitstream/handle/20.500.11912/3126/Trabajo%20de%20grado%20Adriana%20C%C3%B3rdoba.pdf?sequence=1

Cruz, I.M & Puentes, A. (2012). Educational innovation: Use of ICT teaching Basic Mathematics. EDMETIC, Revista de Educacion Mediatica y TIC, 1(2), 127-145. [Accessed on: 23 September 2021] https://www.uco.es/ucopress/ojs/index.php/edmetic/article/view/2855

D'Amore, B. (2006). *Didactica de la matematica*. Bogota: Magisterio. [Accessed: 23 September 2021] https://www.worldcat.org/title/didactica-de-la- matematica/oclc/83095124

De Guzman, M. (1993). Science and Mathematics Teaching. [Accessed: 1 September 2021] https://rieoei.org/historico/documentos/rie43a02.pdf

DEL TORO, M. (2006). Modelo de diseno didactico de hiperentornros de aprendizaje desde una concepcion desarrolladora. UCP. [Consultation: 1 September 2021] https://acacia.red/wp-content/uploads/2019/08/Modelo-de-Creaciones-Didacticas- en-Cooperacion.pdf

D^AZ, O. (2012). UCI: la mirada de un decenio. Granma, Edition del 11 de octubre, (p. 3). [Consultation: 1 September 2021] https://www.granma.cu/granmad/2012/10/11/nacional/artic03.html

Duart, J., & Sangra, A. (2010). Learn virtuality. (2nd ed.). Barcelona: Gedisa. Teaching trends virtual education: interpretive approach Purposes Representations 2018, 6 (1), 463505. [Consultation: 1 September 2021] http://revistas.usil.edu.pe/index.php/pyr/article/view/167

Elliot, H. (2005). Guidelines for conducting a Focus group. American Journal For Reserchers, [Accessed November 14, 2021]. https://assessment.trinity.duke.edu/documents/How to Conduct a Focus Group.p df

ER, E. (2009). LIVELMS: A blended e - learning environment, a model proposition for integration of asynchronous and synchronous e - learning. METU. [Consultation: 1 September2021]. https://www.researchgate.net/publication/287331820 LIVELMS ABlendede-LearningEnvironment AModel PropositionforIntegrationofAsynchronousa nd Synchronous e-Learning

Escobar-Perez, J. and Cuervo-Martinez, A. (2008). Validez de contenido y juicio de expertos: una aproximacion a su utilizacion. En Avances en Medicion, 6 (1), 27-36. [Consultation: 1de September2021]

https://www.researchgate.net/publication/302438451 Validez de contenido y juici o deexpertosUnaaproximacionasuutilizacion

FANDOS, M. (2003a). Training based on Information Communication Technologies: Didactic analysis of the teaching-learning process. Publicaciones URV - Espana. [Consultation: 1 September 2021] https://www.tdx.cat/handle/10803/8909

FANDOS, M. (2003b). Training based on Information Communication Technologies: Didactic analysis of the teaching-learning process. Publicaciones URV - Espana. [Consultation: 1 September 2021] https://www.tdx.cat/handle/10803/8909

FANDOS, M. (2009a). Las tecnolog^as de la información y la comunicación en la educacion: un proceso de cambio. Publicaciones URV - Espana. [Accessed on: 1 September 2021].
https://www.researchgate.net/publication/328383755 Lastecnologias de lainfor macionylacomunicacionenla educacion un proceso de cambio

FANDOS, M. (2009b). Las tecnolog^as de la información y la comunicación en la educacion: un proceso de cambio. Publicaciones URV - Espana. [Accessed on: 1 September 2021].
https://www.researchgate.net/publication/328383755 Lastecnologias de lainfor macionylacomunicacionenla educacion un proceso de cambio

FANTINI, A. (2011). Learning quality evaluation: result indicators Students perception. La Habana: XIV Convencion Internacional y Feria Expositiva - Informatica. [Consultation: 1 September 2021] https://bibliotecaopac.unas.edu.pe/cgi-bin/koha/opac-detail.pl?biblionumber=144792

FERNANDEZ, B. and PARRA, I. (2004). Los medios de ensenanza en la tecnolog^a educativa [Digital]. En COLECTIVO DE AUTORES, Tecnologla Educativa. La Habana: CESOFTE- Division de publicaciones por computadora, Doc. 9. [Consultation: 1 September 2021].
https://anmotoristas.org/documentos/contenidos/libro de tendencias docentes.pdf

FERNANDEZ, S. (2005). Diseno curricular de un proyecto para la ensenanza de ingles por television dirigido a adultos en el contexto cubano. IPL. [Accessed 1 September 2021] https://www.yumpu.com/es/document/view/14625625/tesis-g-sonia- fernan-dr-luis-alberto-montero-cabrera-

Flores & Flores (2007). La evaluacion de competencias laborales. Javier Gil Flores Universidad de Sevilla. ISSN: 1139-613X. Faculty of Education. UNED. Education XX1. 10, 83-106. [Consultation: 1 September 2021].
https://www.redalyc.org/pdf/706/70601006.pdf

Freixo, M. J. V. (2011). Scientific Methodology: Fundamentals, Technical Methods. 3.ª ed. Lisboa: Instituto Piaget. [Accessed: 23 September 2021] https://www.amazon.com.br/Metodologia-Cient%C3%ADfica-Fundamentos-

M%C3%A9todos-T%C3%A9cnicas/dp/9896591148
Frias Cabrera, (2007). Conception of the teaching-learning process UPR. Tesis en opcion al Htulo de Dra. en Ciencias Pedagogicas. [Accessed: 23 September 2021] ftp://ceces.upr.edu.cu/reservorio/tesis/
FR^AS, (2008). Didactic conception blended learning teaching process: application strategy Pinar del Rfo University. [Accessed: 23 September 2021] https://core.ac.uk/reader/53027600
GALLEGOS, D. and PENA, A. (2012). Las TIC en geometria, una nueva forma de ensenar. Bogota: Ediciones de la U. [Accessed: 23 September 2021] https://www.amazon.com/GEOMETRIA-LAS-NUEVA-FORMA-ENSE%C3%91AR/dp/9587620232
Gamboa, R. (2007). I use technology to teach mathematics. Magazine Research Notebooks Training Education Mathematics. 2 (3): 11-44. [Accessed: 23 September 2021] https://revistas.ucr.ac.cr/index.php/cifem/article/view/6890
GAO, P., CHOI, D., WONG, A. and WU, J. (2009). Developper une meilleure comprehension de la pedagogie basee sur la technologie. Journal australien Technologie educative. 25(5). [Accessed: 23 September 2021] https://ajet.org.au/index.php/AJET/article/view/1117
Garda, M. and Benitez, A. (2009). Papier outils de calcul resolution de problemes reflexion etudiants Mathematiques, souvenirs Deuxieme congres international Orientation educative professionnelle, 64-71, Mexico 25 a 27 de marzo. [23 September 2021] https://www.redalyc.org/pdf/3735/373534515005.pdf
Gartia-Valcarcel, A. & Gonzalez, A. D. (2011). IKT-IntegrationSchulpraxis Auswahl von Ressourcen zwei Schlusselbereichen: mathematischeSprache. In R. Roig & C. Laneve (Hrsg.), Bildungspraxis Informationsgesellschaft. Innovationdurch Forschung. pädagogische Praxis nella Societa 'Informazione. L'innovazione attraverso la ricerca (pp. 129-144). Alcoy: Marfil. [23 September 2021] https://dialnet.unirioja.es/servlet/articulo?codigo=3737745
GIBBS, A. (1997). Schwerpunktgruppen. Aktuelles Sozialforschung (Vol. 19). [23 September 2021] https://sru.soc.surrey.ac.uk/SRU19.html
GOMEZ, M. (2002). Estudio teorico, desarrollo, implementacion y evaluation de un entorno de ensenanza colaborativa con soporte informatico. UC. [23 September 2021] https://eprints.ucm.es/id/eprint/4755/
GONZALEZ, A., RECAREY, S. and ADDINE, F. (2004a). The teaching-learning process: challenge for educational change. ADDINE. Didactics. Practical theory. La Habana: Pueblo y Education, (pp. 43 - 65). [23 September 2021] https://profesorailianartiles.files.wordpress.com/2013/03/caracterizacic3b3n-del- proceso-de-ensec3b1anza-aprendizaje.pdf
GONZALEZ, A., RECAREY, S. and ADDINE, F. (2004b). The teaching-learning process: challenge for educational change. ADDINE. Didactics. Practical theory. La Habana: Pueblo y Educacion, (pp. 66 - 84). [23 September 2021] https://www.redalyc.org/pdf/447/44740210.pdf

Gonzalez, F. E. (2000). Lateinamerikanische AgendaforschungMathematik Bildung XXI Jahrhundert. Mathematikunterricht, 12, 107-128. Mexico: Grupo Editorial Iberoamerica, S.A., de C.V [23 September 2021] https://core.ac.uk/download/pdf/154339148.pdf

GONZALEZ, V. (1986). Teaching media practical theory. La Habana: Pueblo y Educacion. [23 September 2021] https://www.estantevirtual.com.br/livros-universo2/vicente- gonzalez-castro-teoria-y-practica-de-los-medios-de-ensenanza 2572865555?show suggestion=0

Gordillo, J. (2017). Utilisation TIC incidence domaines d'apprentissage volumes corps geometriques eleves dixiemeannee formation generale baseunite d'enseignement. Ibarra, Ecuador: Universidad Tecnica del Norte. [23 September 2021] http://repositorio.utn.edu.ec/handle/123456789/6710

GUIZA, M. (2011). Trabajo colaborativo en la web: Entorno Virtual de Autogestion para docentes. UIB.

GUNAWARDEMA, C., LOWE, C. and ANDERSON, T. (1997). Analyse einer globalen Online- Debatte Entwicklung eines Interaktionsanalysemodells Untersuchung sozialen Konstruktion von Wissen Computerkonferenzen. Journal Educational Computing Research, (Vol. 17). [Accessed: 23 September 2021] https://www.researchgate.net/publication/238988854 Analysis of A Global Online Deb ateand The Developmentof anInteractionAnalysis Model for Examining Social Co nstruction of Knowledge in Computer Conferencing.

HENRI, F. (1992). Formation a distance et teleconference assistee par ordinateur: Interactivite, quasi - interactivite, ou monologue? Journal of Distance Education, (Vol. 7, n°. 1). [Consulted on: 1 September 2021]. https://link.springer.com/chapter/10.1007/978-3-642-85098-1 8

HERNANDEZ, R., FERNANDEZ, C. and BAPTISTA, P. (2006). Metodolgia de la investigacion, 4ta edicion. Mexico D.F.: McGraw - Hill. [Accessed: 1 September 2021]. http://observatorio.epacartagena.gov.co/wp- content/uploads/2017/08/metodologia-de-la-investigacion-sexta- edicion.compressed.pdf

HERNANDEZ, S. (2010). Methodische Konzeption Gestaltung von Aufgaben fur studentische Lernen Stadtverwaltung Havanna Agrarian University. UA. [Accessed: 1 September 2021]. https://repositorio.uci.cu/jspui/handle/ident/4588

HERRERA, B. M. (2006a). "Las fuentes del aprendizaje en ambientes virtuales educativos", Revista Iberoamericana de Educacion, ISSN:1681-5653, en http://www.campus-oei.org/revista/de los lectores/352Herrera.PDF, [Consulted on: 1 September 2021]. https://www.redalyc.org/pdf/340/34003507.pdf

HERRERA, B. (2006b). "Las Nuevas Tecnolog^as en el aprendizaje constructivo", Revista Iberoamericana de Educacion, ISSN:1681-5653 [Consulted on: 1 September 2021]. https://doi.org/10.35362/rie3433056

Hidalgo, S., Marato, A., & Palacios, A. (2004). Why they reject mathematics? Multivariate Evolutionary Analysis Mathematical Relevant Attitudes. Revista de educacion (334), 75-95.

[Accessed: 1 September 2021]. https://dialnet.unirioja.es/servlet/articulo?codigo=963460

Hill, M., & Hill, A. (2005). Questionnaire d'enquete. Lisbonne : Edi^oes SHabo.

Horruitiner, P. (2006). *La Universidad Cubana: El modelo de formation*. La Habana: Felix Varela. [Consulted on: 1 September 2021]. https://www.proquest.com/docview/1152019349?pqorigsite=gscholar&fromopenview=tr eu

IMS. Learning Design Information Model (2003). - IMS Global Learning Consortium, Inc. Inc. [en lines]. [Consulted: 1 September 2021]. Available at: http://www.imsglobal.org/learningdesign/index.html

Inacio, R. (2006). Comunidade Virtual Aprendizagem mathematiques - Uma Experiencia avec eleves de 10e annee d'ecole. UL. [Consulted: 1 September 2021]. http://drle.ie.ulisboa.pt/uploads/publications/d38fcb7f975d68a253f44b7a2e116ed1.2006.pdf

INECSE (2005). PISA 2003. Pruebas de matematicas y de solution de problemas. Madrid: Ministerio de Education y Ciencia. [Consulted on: 1 September 2021]. https://www.mecd.gob.es/dctm/ievaluacion/internacional/pisa2003liberados.pdf?documentId=0901e72b801106c6

Infante Quinteiro, P. and Logreira, H. C. (2010). Integration technology education mathematics, 9 (1), 33-46. Universidad Privada Dr. Rafael Belloso Chatin Zulia, Venezuela. [Accessed: 1 September 2021]. https://www.redalyc.org/pdf/784/78415022003.pdf

ISEI-IVEI Basque Institute Educational Research Evaluation (2004). First Evaluation Report PISA 2003, [Consulted on: 1 September 2021]. https://isei- ivei.hezkuntza.net/es/pisa2003

IZQUIERDO, J. and PARDO, M. (2005). Dynamic educational teaching process higher education, employment Information Communication Technologies. University Pedagogy Magazine. Vol. X, n°. 5. [Consulted on: 1 de septiembre 2021]. https://go.gale.com/ps/i.do?id=GALE%7CA466783808&sid=googleScholar&v=2.1&it=r&linkaccess=abs&issn=16094808&p=IFME&sw=w&userGroupName=anon%7E99ee1a7c

JAMES - GORDON, Y. (20 07). A cadre facilitant e-learning efficace ingenierie Environnements developpement. UW. [Consulted: 1 September 2021]. https://pugwash.lib.warwick.ac.uk/record=b2234589~S15

Jaramillo, I. (2007). Strategies visuelles appliquees impact d'acquisition en ligne apprentissage. Praxis Pedagogica. 8, 171-188. [Accessed: 1 September 2021]. https://revistas.uniminuto.edu/index.php/praxis/article/view/960

JARVELA, S. and HAKKINEN, P. (2002). Web - based cases in teaching and learning - The quality of discussions and a stage of perspective taking in asynchronous communications. Interactive Learning Environments [en lfnea], (Vol. 10, n°. 1). [Consulted on: 1 September 2021]. Available at in: https://www.researchgate.net/publication/261691806 WebbasedCasesinTeachingandLearningtheQualityofDiscussionsandaStageofPerspectiveTakinginAsynchron ousCommunication John, B. W. (1980). Latin American Journal Psychology, 12 (3), 540- 541. Konrad Lorenz University Foundation Bogota, Colombia. [Accessed: 23 September 2021] https://www.redalyc.org/pdf/805/80512314.pdf

JOHNSON, R. B., ONWUEGBUZIE, A. J., TURNER, L. A. (2007). Toward a definition of mixed methods research. Journal of Mixed Methods Research, 2 (1), 112-133.

Jonassen, (2007). Computers, Cognitive Tools: developing Cognitive Thinking in Schools. Porto-Portugal: Porto Editora. Cole^ao Ciencias da Educa^ao Seculo XXI, n° 23. [Accessed on: 23 September 2021] https://www.portoeditora.pt/produtos/ficha/computadores-ferramentas- cognitive/183669

Keyler, R.V., Juan, M. P. F., Geisi, T. G. (2018). Implementation of virtual environment teaching tool strengthen teaching-learning process. ISSN 2077-2874 Edumecentro (vol.10 n°.4) oct.-dic. [Accessed: 23 September 2021] http://scielo.sld.cu/scielo.php?script=sci abstract&pid=S207728742018000400004&lng= es&nrm=iso

Lacerda, C.B.F. (2007). Dizem / sentem alunos participantes experience inclusao school student surdo. Revista Brasileira de Educa^ao Especial,13(2), 257-280. [Accessed on: 23de September2021] https://www.scielo.br/j/rbee/a/s6JWTqnb95kYHy38HY6SXLb/abstract/?lang=pt

LAKKALA, M. (2010). Wie die Gestaltung Bildungseinrichtungen kollaborativeForschung fordert Padagogische Infrastrukturen Technologie verbesserte progressive Forschung. UH. [Accessed: 23 September 2021] https://www.researchgate.net/publication/47931424 How to design educational settin gstopromote collaborativeinquiry Pedagogical infrastructures for technology- enhancedprogressiveinquiry

Lara, L. (2001). The dilemma theories teaching learning. Revista Cientffica de Comunicacion y Education, 17, 133-136. [Accessed: 23 September 2021] https://www.redalyc.org/pdf/158/1580170.pdf

LEE, T. (2011). Uberdenken der Verbundenheit: Untersuchung des professionellen Lernens von Access-Lehrern in der Region und im fernen Westaustralien. CU. [Accessed: 23 September 2021] https://espace.curtin.edu.au/handle/20.500.11937/246

Lima, M. S. (2005). Padagogische Mediation nutzt Informations- und Kommunikationstechnologien (IKT). Course 67. Pedagog^a. La Habana. [Accessed on: 23September 2021]. https://www.researchgate.net/publication/313908634 Pedagogical mediation through information and communication technologies and their impact on mathematics teaching in the area of Pococi and Guacimo

Lombillo, I. (2011). Progressive integrated methodological strategy teaching aids teaching means *Universidad* Agraria Habana. [Accessed: 23 September 2021] http://scielo.sld.cu/pdf/rces/v37n3/0257-4314-rces-37-03-e12.pdf

Lopez, A. (2005). Apport du developpement competences "Diagnostic Pathologique" Carriere Medecine Veterinaire. (CREA- CUJAE). [Accessed: 23 September 2021] https://www.proquest.com/docview/1509202387

Lopez, A. (2007). Methodologie d'assimilation Technologies de l'information Communication processus d'enseignement-apprentissage Sciences techniques agricoles. *Revista Ciencias Tecnicas Agropecuarias*, 2(16), 63-68. [Accessed: 23 September 2021] https://www.redalyc.org/articulo.oa?id=93216215

Lopez, A. M. & Gomes, M. J. (2007). Environments Virtual Learning Context Face-to-face teaching: Reflective approach. Challenges - V Conference 292 International Technologies Information Communication Education, Congress, Braga - UM, Ministerio Ciencia, Tecnologia Ensino Superior. [Accessed on: 23 September 2021] https://repositorium.sdum.uminho.pt/bitstream/1822/7098/1/Challenges07-AML- MJG.pdf

LOPEZ, A. and GONZALEZ, V. (2002). Iadov-Technik. bewerbung studienzufriedenheit schulerklassen sportunterricht. Revista Digital EFDeportes.com. 8 (3). [Consulted on: 1 September 2021]. Available at:
https://www.efdeportes.com/efd47/iadov.htm

LOPEZ, F. (2002). El analisis de contenido como metodo de investigacion. Revista de educacion [en lfnea], XXI (4). [Consulted on: 1 September 2021]. Available at: https://dialnet.unirioja.es/servlet/articulo?codigo=309707

Marchesi, Coll and Palacios (1990). Desarrollo psicologico y educacion. Segunda edicion, Alianza Editorial, S. A. Madrid, 2014 Juan Ignacio Luca de Tena, 15. 28027 Madrid ISBN: 978-84-206-8777-3 Edicion en version digital 2014. [Accessed: 23 September 2021] https://dialnet.unirioja.es/servlet/libro?codigo=9974

Marianela, D. F. A. S. G. (s.d). Creative didactic strategies virtual learning environments. Revista Electronica publicada por el IIEUCR ISSN 1409-4703 [Accessed on: 23 September 2021] https://www.researchgate.net/publication/28319119 Estrategias didacticas creativ asenentornos virtualesparael aprendizaje

Martin Arribas, M. C. (2004). Diseno y validacion de cuestionarios. En Matronas Profesion, 5(17), 23-29 [Accessed: 14 November 2021].
https://enferpro.com/documentos/validacion cuestionarios.pdf

Martin Cantoral, (2001). Ensenanza de las matematicas en la educacion superior. Revista Electronica Sinectica (19), 3-27. [Accessed: 23 September 2021] https://www.researchgate.net/publication/44566483 Ensenanzadelamatematica enlaeducacionsuperior

Martin, W. (2000). Effets durables technologies graphiques integrees precalcul mathematiques. E. Dubinsky; A. Schoenfeld; J. Kaput (ed.). CBMS publie l'enseignement des mathematiques. Association mathematique d'Amerique. Washington, D. C. (8), 154-187. PND (2013-2018). [Accessed: 23 September 2021] https://eric.ed.gov/?id=ED390699

Martinez - Padron, O. (2003). Domaine affectif Enseignement des Mathematiques : Aspects Theorico-Referentiels Rencontres Edumatiques. Poste d'ascension affiche. Universite

Pedagogique Experimentale Libertador, Institut Pedagogique, Turmero. matematicas, Revista Iberoamericana de Educacion, 40 (6), 1-11, (2006). [Accessed on: 23 September 2021] http://funes.uniandes.edu.co/15029/1/Mart%C3%ADnez2007Semblanzas.pdf

Matos, (2010). Principes directeurs l'apprentissage Scenarios Learning. Projet Learn Learning : Mathematiques, Society Technology. Lisbonne. [Accessed on: 23 September 2021] https://www.researchgate.net/figure/Figura-1-Principios- orientadores-para-o-desenho-de-cenarios-de-aprendizagem-Matos-2014 figl 329217909

MEN. (2016). Matriz de Referencia matematicas. Siempre D^a-E la ruta hacia la excelencia educativa. Ministerio de Educacion Nacional. [Consulted on: 1 September 2021] .https://www.monografias.com/trabajos38/investigacion- cualitativa/investigacion-cualitativa2.shtml

Mendoza, R. (2006). Investigacion cualitativa y cuantitativa. Diferencias y limitaciones. En: [Consulted on: 1 September 2021]. https://recursos.salonesvirtuales.com/assets/bloques/investigacionDIFERENY LIMI TACIONES.pdf

MILL, D. and PIMENTEL, N. (2010). L'enseignement a distance : defis contemporains. [Consulted on: 1 September 2021]. https://artesanatoeducacional.com.br/produto/educacao-a-distancia-desafios-contemporaneos/

MONDEJAR, J. A., MONDEJAR, J. and VARGAS, M. (2007). Universites virtuelles d'enseignement face: experience l'Universite de Castilla Mancha . (Vol. 10, n°. 2). [Accessed on: 1 September 2021] .https://www.redalyc.org/pdf/3314/331427207010.pdf

Mosquera, J. (2006). Arquitectura y desarrollo. Revista cientffica UNET, 18(2), 46-47. ISBN 1316-869X11C. [Accessed: 1 September 2021]. https://www.researchgate.net/publication/236246960 ARCHITECTURE ANDDEV ELOPMENT

NIKOLAIDOU, M., SOFIANOPOULOU, C.R., ALEXOPOULOU, N. and ABELIOTIS, K. (2009). Exploring a blended learning ecosystem in the academic environment. Rome, Italy: IADIS. [Accessed: 1 September 2021]. http://web.ebscohost.com/ehost/pdfviewer/pdfviewer?sid=1845e171-474f-4c53-%208cd8-%20ecf161683840%40sessionmgr4&vid=1&hid=9

NISANCI, M. (2005). Apprendre l'enseignement superieurfeuille route etablissements d'enseignement superieur turcs leurs efforts cours ligne. MTEU. [Accessed: 1 September 2021]. http://etd.lib.metu.edu.tr/upload/12606254/index.pdf

NOCEDO, I. and ABREU, E. (1984). Metodolog^a de la investigacion pedagogica y psicologica. La Habana: Pueblo y Educacion. [Consulted on: 1 September 2021]. https://books.google.co.ao/books/about/Metodologia de la investigacion pedagogi

.html?hl=es&id=-8yxHAAACAAJ&rediresc=y

NUNEZ, J. (1999). La ciencia y la tecnologia como procesos sociales. Lo que la educacion cientffica no debería olvidar. La Habana: Felix Varela. [Consulted on: 1 September2021]. https://www.researchgate.net/publication/328413184_LACIENCIAYLATECNO_LOGIA_AS_SOCIAL_PROCESSES Lo que la educacion cientificanode beriaolvidar

Parnafes, O. and Disessa, A. (2004). Relations representations computationnelles raisonnement, International Journal Computers Mathematical Learning, Kluwer Academic Publishers, (9), 251-280. [Accessed: 1 September 2021]. https://link.springer.com/article/10.1007/s10758-004-3794-7

PAVON, F. (2008). Virtual classrooms teaching Cadiz University. Latin American Magazine Educational Technology. 7(2). [Consulted: 1 September 2021]. Available at: http://www.uh.cu/static/documents/RDA/Aulas%20Virtuales%20Universidad%20Cadiz.pdf

PEREZ, J. (2002). Entwicklung eines digitalen Lernplattformmodells zur Wissensgenerierung. UC. [Accessed: 23 September 2021] https://eprints.ucm.es/id/eprint/4623/

PICKETT, S. and CADENASSO, M. (2002). Das multidimensionale Konzept des Okosystems: Bedeutung, Modell, Metaphor.(Vol. 5). [Accessed: 23 September 2021] Available at: https://link.springer.com/article/10.1007/s10021-001-0051-y

PINUEL, J. (2002). Epistemolog^a, metodolog^a y tecnicas del analisis de contenido. Estudios de sociolinguistica [en linea]. (Vol. 3, n°. 1). [Consulted on: 1 September 2021].https://www.ucm.es/data/cont/docs/268-2013-07-29%20PinuelRaigada_AnalisisContenido_2002_EstudiosSociolinguisticaUVigo.pd f

Pope, C., Ziebeland, S. & Mays, N. (2006). Analysing qualitative data. In: C. Pope, & N. Mays, (eds.). Qualitative research in health care. 3 rd ed., Oxford, Blackwell, (pp. 63- 81). [Accessed: 23 September 2021] https://www.amazon.com/-/es/Catherine-Pope/dp/1405135123

Proenza, Y. and Leyva, L.M. (2006). Reflexions competences d'apprentissage qualite. Mise place classes virtuelles, formation d'etudiants universitaires. Magaziner echerche technologies l'information. 5(9), 47-53. [Accessed: 23 September 2021] https://rieoei.org/RIE/citationstylelanguage/get/harvard-cite-them-right?submissionId=2479

Ramos, C. (2005). Como investigan los sociologos chilenos en los albores del siglo XXI: Paradigmas y herramientas del oficio. Persona y Sociedad 19 (3): 85-119. [Accessed on: 23 September 2021].

http://sociologia.uahurtado.cl/wpcontent/uploads/2012/01/sociologoschilenos.pdf

Rico, P., Lopez, J., Chavez, J., Ruiz, A. and Valera, O. (2002). Padagogisches Projekt Konzeptionelle Rahmenausarbeitung der Padagogischen Theorie. Zentralinstitut Padagogische Wissenschaften. [Accessed: 23 September 2021] http://www.reed-edu.org/wp-content/uploads/2016/12/libro-REED-1.pdf

Rincon, M. (2008). Environnements virtuels commeoutils conseil academique distance. Magazine virtuel Universidad Catolica Norte. [Accessed: 23 September 2021] https://www.researchgate.net/publication/48198452 Losentornos virtualescomo herramientasdeasesoria academica en la modalidad a distancia

Rincon, M. L. (2008). Environnements virtuels outils conseil academique modalite distance. Revista Virtual Universidad Catolica del Norte, num. 25, septiembre-diciembre, ISSN: 0124-5821. [Accessed: 23 September 2021] https://www.redalyc.org/pdf/1942/194215513009.pdf

RODR^GUEZ, M. (2008). Una estrategia para el diseno e implementacion de cursos virtuales de apoyo a la ensenanza semipresencial en la carrera de econom^a de la Universidad de Camaguey. UH. [Accessed: 23 September 2021] https://www.researchgate.net/profile/Milagro-Rodriguez-   Andino/publication/324970908 TesisMilagro/links/5aee2468aca2727bc0050fd2/T esis-Milagro.pdf

Rozo, O., & Perez, V. (2014). Didactics mathematics technologies information communication. Revista De Educacion y Desarrollo Social, 8(2), 60-78. [Accessed on: 23de Septemberde2021] https://revistas.unimilitar.edu.co/index.php/reds/article/view/295

Salinas, J. (2002). ICT means new university. ICT introduction effects improves university teaching. 2nd International Congress University Teaching Innovation. Tarragona, Espana. [Accessed: 23 September 2021] https://www.redalyc.org/pdf/780/78011256001 .pdf

Sanchez Villegas, D. S. (2018). Objetos virtuales de aprendizaje como estrategia didactica de ensenanza aprendizaje en la educacion superior tecnologica. Ambato-Ecuador. [Accessed on: 14 November 2021]. https://repositorio.uta.edu.ec/bitstream/123456789/28124/1/1804326997-Diego-Sebasti%C3%A1 n-S%C3%A1 nchez-Villegas.pdf

SANCHEZ, J., MUNTADAS, M., SANCHEZ, C. and SANCHO, J. (2008). El campus virtual de la Universidad de Barcelona. Modelos de ensenanza y aprendizaje emergentes. Revista Latinoamericana de Tecnolog^a Educativa [en lines]. 7 (2). [Accessed on: 23 Septemberde2021] https://www.researchgate.net/publication/28243415 El Campus Virtual de la Universi daddeBarcelona modelosde ensenanza y aprendizaje emergentes

Santos, E. O. and Okada, A. L. P. A. (2003). Construgao de Ambientes Virtual de Aprendizagem: Por Autorias Plurais e Gratuitas no Ciberespa^o.

SCARDAMALIA, M. (2002). Reflections on the transformation of education for the knowledge age. [Accessed: 23 September 2021] https://www.researchgate.net/publication/28077375 Reflections on the transformation of education for the knowledge age

Serrano, M.S. (1993). Didaktik der Mathematik. Aufsatze: Albacete Education Faculty Magazine, (8), 173-194. [Accessed: 23 September 2021] https://dialnet.unirioja.es/servlet/autor?codigo=1985113

SILVA, J. (2007). Interactions environnement virtuel apprentissage formation continue enseignants education base. UB. [Accessed: 23 September 2021] https://gredos.usal.es/handle/10366/56584

SILVESTRE, M. and ZILBERSTEIN, J. (2003). Vers developpeur didactique. Havane: Educationgens . [Accessed on: 23 September 2021] https://isbn.cloud/9789591307323/hacia-una-didactica-desarrolladora/

SKELTON, D. (2007). Eine Untersuchung Lernumgebungen Blended Delivery Lernen und Klassenzimmer Tertiare Umgebung. Doktorarbeit, Curtin University. [Accessed: 23 September 2021] https://espace.curtin.edu.au/handle/20.500.11937/555

SKINNER, B. F. (1990). Can Psychology Be a Science of Mind? American Psychologist, 45(11), 1206-10. [Accessed: 23 September 2021]. https://www.semanticscholar.org/paper/Can-psychology-be-a-science-of-mind-Skinner/9252422937bbe2b3ecb84f7ff4cd94f00db4bda9

Sunkel, G. (2006). Bildungstechnologien Lateinamerika. Eine Untersuchung Indikatoren. ECLAC. [Accessed: 23 September 2021] https://www.cepal.org/es/publicaciones/6133-tecnologias-la-informacion-la- comunicacion-tic-educacion-america-latina

TEJADA, J. (1999a). El formador ante las tecnologˆas: roles y competencias profesionales, Comunicacion y Pedagogˆa, 158, 17-26. [Accessed: 23 September 2021] https://hsigrist.github.io/TES2016/1999%20comunicacion%20pedagogia-formador%20ante%20las%20tic.pdf

TEJADA, J. (1999b). Didactica-curricular: diseno, desarrollo e innovation curricular, Proyecto de catedra, Dep. Pedagogˆa Aplicada, UAB, doc. Policopiado. [Accessed on: 23 September 2021] https://www.redalyc.org/pdf/567/56711111.pdf

TEJEDOR, J. (2005). Niveaux satisfaction insatisfaction scolaire Environnement Naturel Activites. Application technique Iadov. Revista EFDeportes.com. 8(3). [Accessed: 23 September 2021] https://www.efdeportes.com/efd85/iadov.htm

THORSTEINSSON, G. and DENTON, H. (2008). Die Entwicklung verstandnisvollen Padagogik Verwendung Virtual Reality-Lernumgebung unterstutzt Innovationsausbildung. (Vol. 13, no. 2). [Accessed: 23 September 2021] https://www.routledgehandbooks.com/doi/10.4324/9780203387146.ch33

UDEN, L. and DOMIANI, E. (2007). The future of E - learning: E - learning ecosystem. Cairns, Australia: IEEE - DEST. [Accessed: 23 September 2021] https://ieeexplore.ieee.org/document/4233689

Unigarro, M.A. (2004). Educacion Virtual. Encuentro Formativo en el Ciberespacio. Bucaramanga: UNAB. [Accessed: 23 September 2021] https://books.google.com/books/about/Educaci%C3%B3nvirtual.html?id=C03hWjUL9OAC

Vasquez, J.L. (2002). Matematicas, Ciencia y Tecnologˆa: una relation profunda y duradera. Encuentros multidisciplinares, 4(11): 22-38. [Accessed: 23 September 2021] https://repositorio.uam.es/bitstream/handle/10486/680588/EM 11 3.pdf?sequence =1

VAZQUEZ, E. (2011). Diseno, implementation de un entorno virtual de formation para la ensenanza de la matematica, basado en los estilos de aprendizaje. UNED. [Accessed: 23

September 2021] https://dialnet.unirioja.es/servlet/tesis?codigo=26404

VERA, A. and VILLALON, M. (2005). La triangulation entre metodos cuantitativos y cualitativos en el proceso de investigation. Ciencia & Trabajo [en lines]. (Vol. 7, n°. 16). [Accessed: 23 September 2021] http://www.uprh.edu/elopez/13%20Triangulacion.pdf

VERDECIA, E. (2011). Metodolog^a para la certification formativa de roles desde la practica profesional. UCI. [Accessed: 23 September 2021] https://repositorio.uci.cu/handle/ident/4591

VILLASEVIL, F. (2009). Design and application of a teaching methodology adapted to the EHEA framework for engineering with multimedia support on a virtual platform. Tesis doctoral, Universidad Nacional de Educacion a Distancia. [Accessed: 23 September 2021] https://www.amazon.com/Metodolog%C3%ADa-Docente- adapted-Marco-Ingenier%C3%ADa/dp/3846569100

Vygotsky, S.L. (1987). Histoire developpement fonctions psychiques superieures. Editorial Technique Scientifique, La Habana. [Accessed: 23 September 2021]
https://www.marcialpons.es/libros/historia-del-desarrollo-de-las-funciones- psychicas-superiores/9789505630554/

VYGOTSKY, L. (2001). Pensamiento y lenguaje. En COLECTIVO DE AUTORES (comp.) Obras escogidas - Problemas de la psicolog^a general, Tomo II. Madrid: A. Machado Libros S.A. First part. [Accessed: 23 September 2021]

https://www.redalyc.org/pdf/1390/139039784004.pdf

WILLIAMS, A. and KATZ, L. (2001). The Use of Focus Group Methodology in Education: Some Theoretical and Practical Considerations. International Electronic Journal for Leadership in Learning [en lines]. [Accessed: 23 September 2021]

https://www.researchgate.net/publication/228941039 The Use ofFocusGroupMetho dology in Education Some Theoretical and Practical Considerations 5 3

ZHU, E. (1996). Meaning negotiation, knowledge construction, and mentoring in a distance learning course. Indianapolis, USA: National Convention of the Association for Educational Communications and Technology [en llnea]. [Accessed: 23 September 2021]

https://eric.ed.gov/?id=ED397849

ANNEXES

Annex. 1
SURVEY APPLIED TO MATHEMATICS TEACHERS

Dear teacher, the purpose of this survey is to **find out your idea about the teaching-learning process supported by the implementation of a virtual environment**. However, we ask for your collaboration and that you answer the questions the way you think is right. The survey is anonymous.

Thank you very much for your collaboration.

Age Gender: Male Female

School

1. Periodicity of access to the virtual environment	
Daily Weekly	Monthly None
2. What are your views on the subject of mathematics?	
Very difficult ⊔⊔ Difficult	Acceptable ‖Easy ‖Very Easy ｜⊔⊔ Very Easy ⊔⊔
3. Regarding each of the Mathematics.	In the following statements, please point out the reasons for the students' difficulties in
3.1. Little study.	Never Rarely Sometimes Often Always
3.2. Deficient bases.	1 2 3 4 5
3.3. Lack of motivation for discipline.	
3.4. Inadequate teaching methods used by teachers.	
3.5. No support outside the classroom.	
3.6. Other	
4. How long have you been giving mathematics lessons?	
I have started this 2019 course Started last year 2018 From 2 to 5 years From 6 to 10 years Over	
5. Do you consider VLE as a support tool for students in Mathematics?	
Yes, it's very useful. I don't know about it. I don't use AVA.	
6. Do you like teaching mathematics?	
I like it a lot ⊔⊔ I like it ⌊ I like it a little ⌊ ⌋ I don't like it ｜⊔⊔	

COMPUTER AND INTERNET USE OUTSIDE THE CLASSROOM

7. Do you have a computer at home?	Yes		No
	If not, do not answer question 8.		
8. Is your computer at home connected to the Internet?	Yes		No
9. Do you enjoy using the computer?	I like it very much		I like
	I don't like it much		I don't like

10. In which place and periodicity do you usually (Do you have the possibility to select more than d	use the computer? and an alternative)			
	Never	Rarely	Sometimes	Every day 4
	1	2	3	
At home				
At home with relatives				
At friends' houses				
In public places				
At school:				
Elsewhere(s). Which one(s)?				

11. During the school year, how often and for what purpose do you use the computer?				
To:	Never	Rarely	Sometimes	Always
Elaboration of school work.				
Presentations in PowerPoint.				
Investigate on the internet				
Communicate with friends/colleagues (Messenger/Skype/Facebook,...)				
Access educational sites (educational sites are those that enable you to build up your knowledge)				
Have access to the school website				
Have access to the school's teaching and learning platform "Moodle".				
Swap emails				
Watching movies or listening to music				
Play				

12. Where and how often do you usually access educational websites? (You have the possibility to select more than one option)				
	Never	Rarely	Sometimes	Everyday
At home				
At home with relatives				
At friends' houses				
In public places				
At school				
Elsewhere(s). Which one(s)?				

13. What kind of educational websites do you usually access and how often? (You have the possibility to select more than one option)				
	Never	Rarely	Sometimes	Every day
Information sites				
Educational sites				
Educational games sites				
Blogs				
Wikis				
Webquests				
Other(s) Which one(s)?				

14. For what purpose and how often do you access educational websites? (You may select more than one option)				
	Never	Rarely	Sometimes	Every day
To study				
To clarify doubts				
Answer students' questions				
To elaborate on tasks proposed by teachers				
Out of curiosity / taste				
Other(s) Which one(s)?				

15. Are you registered on the school's Moodle teaching and learning platform?	YesNo If not, go to question 18.
16. How often do you access the platform?	Never accessed after my registration Rarely Weekly Daily

If you never had access to the platform after your registration do not answer the following question.

17. For what purpose and how often do you use the platform? (You may select more than one option)

	Never Rarely Daily Weekly
Access to information	
Access to resources	
Clarify doubts of the students	
Orienting works	
Guiding discussion forums	
Participate in Chats	
Collaborate in the construction of glossaries	
Orienting work on the Wiki	
Other(s). Which one(s)?	

18. Which of the activities below was most beneficial to your teaching and learning process? (You have the possibility to select more than one alternative)

	Never Rarely	Weekly	Daily
Preparation of readings or documents			
Reading materials			
Participation in forums			
Video assistance			
Information videos			
Video tutorials			
Manuals			
Participation in forums			
Elsewhere(s). Which one(s)?			

19. During the orientation of the distance learning tasks, what kind of problems did you have accessing the VLE? (You have the possibility to select more than one alternative)
If you have not had any problems accessing the platform please move on to the next question.

	Never Rarely	Almost always Always
Lack of time		
Difficulty connecting to the Internet		
Slow access to the platform		
Forgotten "username".		
Forgetting the "password		
Other problem(s). Which one(s)?		

9. Do you consider it important to use VLE as a complement to face-to-face teaching, as a way of stimulating and favouring the teaching process?	Not important Important	Not very important Very important
20. Considers that the use of the VLE increased your motivation to develop and build your knowledge and that of the students in relation to the teaching units covered?	No increase Increased	Increased little It has increased a lot
21. Considers that the use of the VLE was important for information sharing and of construgao mnhAririAntn nartilharln?	Not important Important	Not very important Very important
22. In general, do you consider the use of VLE important in the learning process?	Not important Important	Not very important Very important
23. Were techniques and strategies promoted as a teacher/virtual tutor necessary?	Did not promote Promoted	Promoted little It promoted a lot
24. Do you consider that the resource to the communication tools used in the VLE has promoted greater teacher-student interaction; student-student?	Did not promote Promoted	Promoted little It promoted a lot
25. Do you consider that the use of VLE has promoted greater interaction with the contents?	Did not promote Promoted	Promoted little It promoted a lot

Annex. 2

STUDENT SURVEY

Dear student, the purpose of this survey is to know your idea about **the teaching-learning of mathematics with the support of the implementation of a virtual environment.**
However, we ask for your collaboration and that you answer the questions the way you think is right. The survey is anonymous.
Thank you very much for your cooperation.

Age Gender: Male Female

School

1. How much time per week do you dedicate to the study of Mathematics?
Less than 1 hour 1 to 3 hours 3 to 5 hours More than 5 hours
2. What do you think about the subject of Mathematics?
Very difficult O DificiU 1 Acceptable I	Easy I	Very easy 11
3. For each of the statements below, indicate to what are the reasons for your difficulties in mathematics. Never Rarely Sometimes Often Always

	1	2	3	4	5
3.1 Little study					
3.2. deficient bases.					
3.3 Absence of motivation by the discipline .					
Inappropriate use of teaching methods by teachers.					
3.5. Absence of out-of-class support					
3.6. Other					

4. What grade did you achieve in Mathematics in the previous school year?
IO   2030   40   50
5. The subject of Mathematics c   onsidered u im student:
Very weak L |Frac   o O   Reasonable   Good   Very good
6. Do you like Mathematics?
I like it a lot ||| I like it 111   I don't like it much   I don't like

COMPUTER AND INTERNET USE OUTSIDE THE CLASSROOM

7. Do you have a computer at home?	Yes	No
	If not, do not answer question 8.	
8. ~~Does~~ your computer at ~~home have an Internet connection?~~	Yes	No
9. Do you enjoy using the computer?	I like it very much	I like
	I don't like it much	I don't like

10. Where and how often do you usually use the computer? (You have the opportunity to select more than one option)

	Never	Rarely 1 2	Sometimes 3	Every day 4
At home				
At home with relatives				
At friends' houses				
In public places				
At school:				
In (an)other location(s). Which one(s)?				

11. During the school year, how often do you use the computer? (You have the opportunity to select more than one option)

To:	Never	Rarely	Sometimes	Always
Preparing school work				
Making PowerPoint presentations				
Search the internet				
Communicate with friends/colleagues (Messenger/Skype/Facebook,...)				
Having access to educational sites (educational sites are those that allow you to build up your knowledge)				
Access to the school website				
Access the school's teaching and learning platform "Moodle".				
Swap emails				
Watching movies or listening to music				
Play				

12. Where and how often do you usually access educational websites? (You have the possibility to select more than one option)

	Never	Rarely	Sometimes	Everyday
At home				
At home with relatives				
At friends' houses				
In public places				
At school				
In (an)other location(s). Which one(s)?				

13. What kind of educational websites do you usually access and how often? (You have the possibility to select more than one option)

	Never	Rarely	Sometimes	Every day
Information sites				
Educational games sites				
Blogs				
Wikis				
Webquests				
Other(s) Which one(s)?				

14. For what purpose and how often do you access educational websites? (You may select more than one option)

	Never	Rarely	Sometimes	Every day
To study				
To clarify doubts				
To do tasks proposed by teachers				
Out of curiosity / taste				
Other(s) Which one(s)?				

15. Are you registered on the school's Moodle teaching and learning platform?	Yes	No
	If not, go to question 18.	
16. How often do you access the platform?	Never accessed after my registration	Weekly
		Rarely
		Daily

If you have never logged on to the platform after your registration do not answer the following question.

17. For what purpose(s) and how often do you use the platform? (You have the possibility to select more than one

option)

	Never	Rarely	Daily	Weekly
Access information				
Accessing resources				
Clarify doubts				
Handing in assignments				
Participate in discussion forums				
Participate in Chats				
Collaborate in the construction of glossaries				
Doing work on Wiki				
other(s). Which one(s)?				

18. During the execution of the distance learning tasks, how often did you access the virtual learning environment? (You have the possibility to select more than one alternative)

	Never	Rarely	Weekly	Daily
At home				
At home with relatives				
At friends' houses				
In public places				
At school:				
In (an)other location(s). Which one(s)?				

19. During the distance learning tasks, what kind of problems did you have accessing the VLE? (You have the possibility to select more than one alternative)
If you have not had any problems accessing the platform go to the next question.

	Never	Rarely	Almost always	Always
Lack of time				
Difficulty connecting to the Internet				
Slow access to the platform				
Forgotten "username".				
Forgetting the "password				
Other problem(s). Which one (s)?				

9. Do you consider it important to use VLE as a complement to face-to-face teaching, as a way of stimulating and favouring the teaching process?	Not important Not very important Important Very important
20. Do you consider that using the VLE has increased your motivation to develop and build up your knowledge in relation to the teaching units taught?	Did not increase Increased Increased a little Increased a lot
21. Do you consider that the use of VLE was important for the sharing of information and the construction of shared knowledge?	Not important Important Not very important Very important
22. In general, it considers the use of VLE in the learning process?	Nothing important Not very important Important Very Important
23. Do you consider that the use of the communication tools used in the VLE has promoted greater interaction between student(s) /student(s)?	Did not promote Promoted little Promoted It promoted a lot
24. Do you consider that the use of the communication tools used in the VLE has promoted greater teacher-student interaction?	Did not promote Promoted little Promoted It promoted a lot
25. Do you consider that the VLE support promoted greater interaction with the contents?	Did not promote Promoted little Promoted It promoted a lot

Annex 3: Operationalization of the variable Didactic Design of a VLE.

Annex 3.1: Dimension teachers involved in VLE design.

Indicator	Rating Scale
Number of teachers who have obtained pedagogical preparation to didactically design the VLE.	High: between 70% and 100%. Medium: from 50% to 69%. Low: less than 50%.
Number of teachers who have collaborated on the didactic design of the VLE.	High: between 70% and 100%. Medium: from 50% to 69%. Low: less than 50%.
The teachers' degree of knowledge about the components to be taken into account in the didactic design of a VLE.	High: more than 60% concerns five or more of the components. Medium: more than 60% concerns three or four of the components. Low: the above conditions are not verified.
Degree of knowledge that teachers state about how to design the use of VLE to integrate the EAP.	High: more than 60% concerns five or more design models. Medium: more than 60% concern three or four design models. Low: when the above conditions are not verified.
Teachers' satisfaction with the VLE teaching project.	High: there are no precise disclosures to redesign the AVA. Medium: less than 30% have isolated revelations. Low: over 60% have revelations.
Figure of teachers who use the VLE in the PEA.	High: between 70% and 100% claim to use it a lot. Medium: between 70% and 100% claim to use it little. Low: between 70% and 100% claim never to use it.
Number of teachers who have obtained pedagogical preparation to didactically design the VLE.	High: between 70% and 100% claim to use it a lot. Medium: between 70% and 100% claim to use it little. Low: between 70% and 100% claim never to use it.
The objectives that teachers pursue when using VLE in the EAP	Propagation: over 60% use it for content spreading and task gathering. Exchange: over 60% use it for information exchange. Collaboration: more than 60% use it for collaboration.
Frequency of use of the VLE by the teachers of the PEA.	High: more than 60% use it at least once a day. Average: more than 60% use it at least two to three times a week. Low: more than 60% use it at least once a week.
Teachers' level of knowledge about the didactic design of a VLE.	High: over 60% claim to have high knowledge. Medium: more than 60% claim to have medium knowledge Low: when the above conditions are not verified

Appendix 3.2: Dimension students using VLE.

Indicator	Scale of Classification
Student satisfaction with VLE design.	High: no revelations of dissatisfaction. Medium: less than 30% have isolated findings of inconsistency. Low: over 60% reveal dissent.
Student satisfaction with the possibilities offered by the ALE in the EAP.	Halt: there are no incosistence revelations. Medium: less than 30% have dissenting disclosures. Low: over 60% reveal dissent.
Objectives that students seek when using the VLE of the subject.	Propagation: over 60% use it for spreading content and delivering homework. Exchange: over 60% use it for information exchange. Collaboration: over 60% use it for collaboration.

Annex 3.3: Teaching design dimension of the VLE.

Annex 3.3.1: Semantic subdimensioning.

Indicator	Scale of Classification
Degree of coherence of the system didactic system implemented in the VLE with the subject programme.	High: between 70% and 100% of the SAP components that appear in the VLE. Medium: between 50% and 69% of the ADP components that appear in the VLE. Low: at least 50% of the PEA components that arise in the AVA.
Degree of integration with the PEA through the AVA.	High: between 70% and 100% of the tasks involve working with information. Medium: between 50% and 69% of tasks involve working with information. Low: less than 50% of the tasks involve working with information.

Annex 3.3.2: Technological subdimension.

Indicator	Rating Scale
Degree of coherence of the VLE technological context with the analytical programme of the subject.	High: between 70% and 100% of the AVA instruments contest to an end of the analytical plan. Medium: 50% and 69% of the AVA instruments object to an end to the analytical plan. Low: less than 50% of AVA instruments object to an end to the analytical plan.
Degree of identification by the participants of the VLE technology system and its correlation.	High: there are no revelations of lack of knowledge of instruments and what they are used for. Medium: less than 30% know the instruments and what they are used for Low: when the above conditions are not verified.
Degree of incorporation of the technology stage for AVA.	High: between 70% and 100% of instruments provide VLE access. Medium: between 50% and 69% of instruments provide VLE access. Low: less than 50% of instruments provide VLE access.

Annex 3.3.3: Practical subdimension.

Indicator	Rating Scale
Degree of coherence between the didactic strategy to develop the semi-attendance SAP of the subject and the design that supports it in the VLE.	High: between 70% and 100% of the VLE tasks challenge the teaching strategy. Medium: between 50% and 69% of the AVA tasks challenge the didactic strategy. Low: less than 50% of the AVA tasks challenge the didactic strategy.

Annex 3.3.4: Spatial subdimension.

Indicator	Rating Scale
Degree of coherence of the VLE virtual space system with the SAP scenarios.	High: between 70% and 100% of the SAP scenarios are matched in the VLE. Medium: between 50% and 69% of the SAP scenarios are matched in the VLE. Low: less than 50% of the PEA scenarios are matched in the VLE.
Degree of definition of virtual spaces for function interaction.	High: there is more than 1 space. Medium: there is at least 1 space. Low: there are no virtual spaces.

Indicator	Rating Scale
Degree of definition of virtual spaces for function interaction.	High: more than 1 space or instrument exists. Medium: there is at least 1 space or instrument. Low: there are no spaces or instruments.
Degree of definition of virtual space to individualize teaching - learning.	High: more than 1 space or instrument exists. Middle: there is at least 1 space or tool. Low: there are no spaces or tools.
Degree of definition of virtual space for intercommunication with participants outside the institution.	High: the spaces effect ends that complement each other and are not juxtaposed. Medium: there are several spaces to achieve a similar end and they overlap. Low: there are spaces that do not contest any purpose and, in the meantime, are not linked.
Degree of bonding in the use of the different virtual spaces defined in the VLE.	High: the spaces effect ends that complement each other and are not juxtaposed. Medium: there are several spaces to achieve a similar end and they overlap. Low: there are spaces that do not contest any purpose and are therefore not linked.

Annex 3.3.5: Personal subdimension.

Indicator	Rating Scale
Degree of coherence of the charges in the AVA with those identifiable in the software industry.	High: more than 70% of the Pidsw charges exist in the VLE and can have participation. Medium: between 50% and 69% of the charges that exist in the VLE and can have participation. Low: less than 50% of the charges that exist in the VLE and can have participation.
Degree of coherence of the charges in the VLE with those required for a semi-potential web-based EAP.	High: more than 70% of the SAP roles exist in the VLE and can have participation. Medium: between 50% and 69% of the SAP roles exist in the VLE and can have participation. Low: less than 50% of the PEA charges exist in the VLE and can have participation.

Annex 3.3.6: Management sub-dimension.

Indicator	Rating Scale
Degree of technological management (security and access) that the VLE has.	High: the whole VLE and more than 70% of the activities have access control and permissions. Medium: the whole VLE and between 50% and 69% of the activities

	have access control and permissions. Low: the above conditions are not verified.
Level of support and pedagogical technological support provided by the VLE.	High: the pedagogical and technological support and the support procedures are programmed. Medium: the assistance and support procedures are programmed in a pedagogical and technological way. Low: no assistance procedure has been programmed.
Degree of use of international rules of the VLE project and its technological tools.	High: the entire AVA and more than 70% of its components use some international model. Medium: the whole VLE and between 50% and 69% of its components use some international standard. Low: the above conditions are not verified.

Annex 3.4: Production process dimension - overcoming AVA.

Indicator	Rating Scale
Degree of use of a production process - overcoming AVA.	High: and defined and used. Medium: is defined, but no more than 50% of the defined achievements are realised. Low: not defined and only his isolated achievements are realised.
Level of use of a diagnostic to teachers for incorporation into the production process - progress.	High: and defined and used. Medium: is defined, but not used. Bass: not defined.
Level of linkage of the VLE production process with the teachers' progress.	High: more than 70% of the teachers declare that the achievements of the production process of a VLE are linked to its progress. Medium: between 50% and 69% of the teachers declare that the actions of the process of production of a VLE are articulated with its progress. Low: less than 50% of the teachers declare that the achievements of the VLE production process are articulated with its progress.

Annex 4: Operationalisation of the SAP Integration variable.
Annex 4.1: Interaction of the dimensions between the VLE participants.
Annex 4.1.1: Participatory sub-dimensioning.

Indicator	Rating Scale
Media of messages or statements issued by the participant.	High: every day each participant issues a message or statement. Medium: every two days every participant issues a message or statement. Low: every 3 or more days every participant issues a message or statement.

Annex 4.1.2: Interactive subdimension.

Indicator	Rating Scale
cipher of messages or statements linked together forming chains.	High: between 70% and 100% are connected. Medium: between 50% and 69% are connected. Low: less than 50% are connected.

Annex 4.1.3: Functional subdimension.

Indicator	Rating Scale
encryption of messages or instructions whose role is social.	High: more than 30% to collect reflections not related to the subject content, but to determine social exchanges. Medium: between 16% and 30% to gather reflections not related to the subject content, but to determine social exchanges. Low: less than 16% to gather reflections not related to the subject content, but to determine social exchanges.
encryption of messages or instructions whose role is technical.	High: more than 30% for support, interrogation or response to technical issues. Medium: between 16% and 30% for support, questioning or answering technical aspects. Low: less than 16% to assign support, questioning or response to technical aspects.
encryption of messages or statements whose paper organizational.	High: more than 30% should carry out approaches to work organisation in the VLE. Medium: between 16% and 30% to carry out approaches to work organisation in VLE. Low: less than 16% to carry out approaches of work organization in VLE.
encryption of messages or instructions whose role is to support.	High: more than 30% should carry out proposals to support learningmediately using the VLE. Medium: between 16% and 30% to carry out learning support approachesmedimediate to the VLE. Low: less than 16% should carry out proposals to support learning through the VLE.
cipher of messages or instructions whose role is cognitive.	High: more than 30% should expose general knowledge and skills related to learning progression. Intermediate: between 16% and 30% to display general knowledge and skills relating to learning progress. Low: less than 16% should display general knowledge and skills regarding learning progression.
Cipher of messages or instructions whose role is metacognitive.	High: more than 30% must make claims that certify self-control and self-regulation in learning. Medium: between 16% and 30% should undertake approaches that certify self-control and self-regulation in learning. Low: less than 16% should undertake approaches that certify self-control and self-regulation in learning.
Cipher of messages or instructions whose role is reflective.	High: more than 30% for qualification and verification of theoretical positions or learning outcomes that have been socialised. Medium: between 16% and 30% for qualification and verification of theoretical positions or learning outcomes that have been socialised. Low: less than 16% for qualification and verification of theoretical positions or learning outcomes that have been socialised.

Annex 4.1.4: Direction subdimension.

Indicator	Rating Scale
Cifrade messages or instructions whose address is horizontal.	High: more than 40% must assign criteria without processing or appealing directly to the participation of some other subject, but only share ideas. Medium: between 10% and 40% should assign criteria without directly processing or appealing to the participation of any other subject, but only share ideas. Low: less than 10% should award criteria without processing or appealing directly to the participation of any other subject, but only share ideas.
Cipher of messages or statements from whom the	High: more than 40% should constitute communication with others with better/worse and more/less demonstrated performance. Medium: between 10% and 40% should constitute communication with other people with better/worse and more/less demonstrated performance.

Annex 4.1.5: Scope of the sub-dimension.

Indicator	Rating Scale
Cipher of messages or instructions whose scope is partial.	High: more than 40% allude only to academic research activity or work-research activity. Medium: between 10% and 40% refers only to academic activity - research or work activity - research. Low: less than 10% allude only to academic research activity or work research activity.
Cipher of messages or instructions whose scope and reciprocity.	High: more than 40% translates the content of the academic activity - research work activity - research and vice versa. Medium: between 10% and 40% translates the content of the academic-research activity to work-research activity and vice versa. Low: less than 10% translates the content of the academic-research activity to work-research activity and vice versa.

Annex 4.2: Dimension of interactivity of VLE participants with digital tools.

Annex 4.2.1: Participatory sub-dimensioning.

Indicator	Scale of Classification
Average access to digital tools per participant.	High: each participant accesses the VLE tools more than 10 times a day. Medium: Each participant accesses the VLE tools 5 to 10 times a day. Low: Each participant accesses the VLE tools less than 5 times a day.

Annex 4.2.2: Functional subdimension.

Indicator	Rating Scale
Access cipher whose role is to consult the content of the digital tool.	High: more than 50% for consultation of its contents. Medium: between 10% and 50% for consultation of its contents. Low: less than 10% for consultation of its contents.
Access cipher whose role is to contribute to the content of the digital tool.	High: more than 50% has to contribute to the content. Medium: between 10% and 50% must contribute to the content. Low: less than 10% has to contribute to the content.

Appendix 5: Group interview guides.

Appendix 5.1: Teachers' group interview guide.

Objective: To deepen the teachers' collective criteria issued aloud regarding the didactic design of a VLE and its contribution to the integration of the PEA.

Development of the interview:

Messages and communication of the purpose pursued by the interviewer.

- Individual presentations of the interviewees (first and last name, category of teacher, scientific category, postgraduate speciality, experience as a teacher, experience in the teaching discipline)
- Aspects to be addressed in the interview:

1. What do you understand by a VLE?
2. What components should a VLE have for the Mathematics subject? Why?
3. What does the educational design of a VLE depend on or determine to a high degree? Why?
4. Do you think it is necessary to didactically redesign the VLE? Why do you think so?
5. What in particular should be redesigned? Why?
6. Do you consider that the VLE enables the integration of the ASAP for the subject of Mathematics? Why?
7. What kind of VLE activities and tools would make this integration possible? Why?
8. Do you feel qualified to didactically design the VLE and contribute to this integration?

Why?

9. Is your participation in the production of the VLE considered an activity of pedagogical improvement? Why?

10. What activities do you carry out while participating in the production of the VLE?

11. Does the VLE contribute to your pedagogical improvement? Why?

- Review of data collection.
- Conclusion of the interview.

Appendix 5.2: Group interview guide for students.

Objective: To deepen students' collective criteria, expressed aloud, regarding the objectives and functions of VLE use and satisfaction with this use.

Development of the interview:

Exit and communication of the objective pursued by the interviewer.

- Individual presentations of the interviewees (first and last name, type of software development project in which they work - research activity of the subject, subject they study).
- **Aspects to be addressed in the interview:**

1. What do you think an AVA should look like? Why?
2. What criteria do you have in the AVA of the subject you are studying? Why?
3. What do you consider could be modified in the subject's VLE? Why?
4. Do you use VLE to communicate and collaborate with each other and with the teachers in solving professional problems they solve? Why do you use it?
5. Do they use all the tools available in the VLE? Do you know what to use each one of them for?
6. What are the benefits of using these tools? Why?
7. How do you consider that the VLE could contribute to the software development projects they carry out to integrate with the subject they are taking on?

- Review of data collection.
- Conclusion of the interview.

Annex 6: Document analysis guide.

Objective: To analyse the information that appears in the source in relation to the didactic design of the VLE of the subject and its contribution to the integration of the PEA.

FORMAL ASPECTS	
Research registration figure	
Format	
Source	
Year of creation	
Location	
Author(s)	
ASPEC TEOR	:TOS COS
Essential categories that appear in the document directly related to the didactic design of a VLE.	
Essential categories appearing in the document related to the academic environment - academic integration.	
Other categories with some kind of relationship with VLE.	
Judgments, evaluations and criteria of VLE use for academic integration.	
Other comments of interest to Research.	

Annex 7: Composition of focus groups.

Group	Direction level	Representatives	Observances
IPGB teachers' coordinators	Department	12 full time teachers, four (4) with a master's degree in Mathematics, 11 linked to productive projects in the IPGB and one (1) linked to educational innovation projects. Seven (7) with experience in ICT projects in education. One (1) of them manager.	Average years of teaching experience: 5 Average years of project experience: 6 Average years of management experience: 3
Board of Directors Department		10 specialists, all linked to teaching in Subjects of sciences, six (6) teach in the disciplines d; subject Mathematics. Two (2) masters of science in specialties. All with experience in AVA projects Three (3) of them administrators of the IPGB	Average years of teaching experience: 4 Average years of experience in production: 5 Average years of management experience: 4
Department for Mathematics Teaching	IPGB	13 full time teachers, four (4) with master's degrees in science. Seven (7) with experience in projects. Six (6) connected to ICT in education projects and the rest connected to educational innovation projects. 11 of them teaching managers.	Average years of teaching experience: 5 Average years of project experience: 5 Average years of management experience: 4
Board of Directors of the Directorate General of		14 specialists, 12 linked to the teaching of Mathematics, seven (7) linked to ICT in education projects. Three (3) Masters of Science. All of them with experience in educational innovation projects. Six (6) of them administrators of the IPGB.	Average years of teaching experience: 5 Average years of project experience: 6 Average years of management experience: 4

Appendix 8: Topic guide for the focus group workshops.

Theme 1: Evaluation of the principles, dimensions and relations of the didactic design of

AVA for mathematics discipline.

1.1 What do you think about the educational design of the VLE for the subject of Mathematics?

1.2 How do you evaluate the principles, dimensions and relationships of didactic design?

1.3 What criteria deserve the incorporation of VLE demands, the diagnosis of knowledge and motivations for teachers and students to determine the fundamentals and principles of ICT in an educational context?

Theme 2: Evaluation of the implementation guide of the VLE didactic project for the Mathematics subject at the IPGB.

2.1 What do you think of the implementation guide?

2.2 What criteria does the organisation of the implementation guide deserve from the perspectives of VLE in the educational context and VLE as an educational context?

2.3 What would you add or exclude from the implementation guide to favour its use in teacher improvement?

2.4 What are the elements for your consideration that may limit the generalisability of the implementation guide in the IPGB?

Theme 3: Evaluation of the educational design of the VLE for the subject of Mathematics in relation to its contribution to the integration of the PEA and the IPGB.

3.1 Do you think it is possible that the educational design of the VLE for the subject of mathematics contributes to the integration of the SAP and the IPGB?

3.2 What would you add or exclude from the educational design of the VLE for Mathematics to increase its contribution to the integration of the PEA and IPGB?

Topic 4: Coherence between the didactic design of the VLE for the subject of Mathematics and the actual conditions at the IPGB, to contribute to the integration of the SAP.

4.1 Do you consider that the didactic design of the VLE for the subject of mathematics contains the essential aspects to be considered in the IPGB in order to contribute to the integration of the SAP of the subject of mathematics?

4.2 What are the elements for your consideration that may limit the widespread implementation of the VLE didactic design in the IPGB?

Annex 9: Operational criteria for focus groups.

Operational criteria	Rating scale
Unanimity of criteria.	It was considered when all focus groups agreed on the answer, and the opinions given within each group had consensus.
Most of the criteria.	It was considered when three of the groups and more than half plus one of the members of each group agreed with the answer.
Minority of criteria.	When a single group and less than half of the members of each group matched the same type of response.

Appendix 10: Final report of the focus group workshops.

This report summarises the key issues raised by participants in the focus groups and discusses the operational criteria adopted for each of the questions that guided the workshops.

Theme 1: Evaluation of the principles, dimensions and relations of the didactic design of

AVA for mathematics discipline.

1.1 What do you think about the educational design of the VLE for the subject of Mathematics?

The focus groups agreed on the relevance of didactic design, in line with the predominant need in the IPGB to conduct educational research that scientifically supports didactic design and provides the basis for methodological work, especially in the mathematics specialty departments. The expression that exists in all its components was highlighted as relevant, which clearly indicates the intention to integrate it into teaching. It is recognised that it is comprehensible and well represented schematically.

Operational criteria: unanimity of criteria, given by the consensus existing in the four focus groups and absence of unfavourable or discrepant criteria.

1.2 How do you evaluate the principles, dimensions and relates of didactic design?

The focus groups evaluated the principles, dimensions and relates of the didactic design as correct, except for a minority of groups 2 and 4, who considered that the incorporation in the spatial dimension of a space as a homologous of the physical space where the student also learns should be valued and fill a gap that today exists in the industrial projects of virtual spaces where collective activity takes place. Regarding the principles, group 3 proposed that the support principle should be valued, because if this support is always provided in technological terms to the participants, the ICT learning curve is slower.

Operational criteria: most criteria, as three of the four focus groups had consensus from their members. However, according to the criteria of the members of groups 2 and 4 on the spatial dimension, they did not represent more than 50% plus one in any case.

Agreement 1: to evaluate the incorporation in the spatial dimension of a space for the Mathematics team.

Agreement 2: review the conception of the principle of backing and support.

1.3 What criteria merit the incorporation of demands, knowledge diagnostics and motivations for teachers and students to determine the foundations and principles of VLE as an educational context?

The incorporation of the demands and diagnoses of teachers and students to determine the didactic design of VLE as an educational context, the focus groups agreed to evaluate as positive. They stated that the evaluation of the former would allow one to consider its restrictions and limitations. Regarding the teachers' diagnostics, in the knowledge component, most of the groups 2 and 4 considered that it should include the diagnosis of knowledge related to the discipline and not only to ICT and VLE, which facilitated the articulation in the methodological work, the pedagogical, technological and content components of the respective discipline.

Theme 2: Evaluation of the implementation guide of the VLE didactic project for the Mathematics subject at the IPGB.

2.1 What do you think of the implementation guide?

The implementation guide was evaluated by the focus groups in a very positive way, essentially its viability in incorporating the improvement of teachers and experts for the didactic design and the use of a VLE. However, in groups 1, 3 and 4, criteria were issued that included support and clarifications about the interoperability of the tools used in the two processes in the implementation - assembly phase. Operational criteria: unanimity of criteria, motivated by the consensus existing in the four focus groups and by the absence of unfavourable or discrepant criteria in most groups.

2.2 What criteria does the organisation of the implementation guide deserve from the perspectives of VLE in the educational context and VLE as an educational context?

A consensus was reached regarding the criteria of the different focus groups. The issuing of criteria was favourable in terms of the coincidence between the didactic design and the stages of the guide, which allow for quick orientation of teachers in its use. They argued that most didactic projects become unusable because the implementation guides differ too much from these, which does not occur in the case analysed. However, in groups 1 and 4, several members expressed their opinion on whether the content of statistical analysis is not precise as part of the improvement of teachers in the last phase.

Operational criteria: unanimity of criteria, motivated by the consensus existing in the four focus groups and by the absence of unfavourable or discrepant criteria in most groups.

Agreement 4: assess the inclusion of statistical analysis issues in the teacher improvement in the last phase of the implementation guide.

2.3 What would you add or exclude from the implementation guide to favour its use for the improvement of teachers?

In relation to the approaches taken in the previous questions, the focus groups reiterated their consensus.

Operational criteria: unanimity of criteria, motivated by the consensus existing in the four focus groups and the absence of unfavourable or discrepant criteria in most groups. Agreement 4 previously signed and ratified.

2.4 What are the elements for your consideration that may limit the generalisability of the implementation guide in the IPGB?

Groups 1, 2 and 4 expressed concern about the length of time that the experts have to actively participate in the aids contained in the guide. They decided to evaluate the use of other ways of carrying out these aids in the case of these participants. It was then clarified to them that the forms, methods and means by which the aids of the guide are carried out will depend on the specific conditions in which they are executed, but it does not imply that the VLE itself in construction can serve as a space for the execution of the aids, if necessary, and to reduce the space-time barriers of the participating teachers.

Operational criteria: most of the criteria, considering that three of the four focus groups had the consensus of their members. Agreement 5: include in the explanation of the implementation guide the ways, methods and means for its execution.

Theme 3: Evaluation of the educational design of the VLE for the subject of Mathematics in relation to its contribution to the integration of the PEA.

3.1 Do you think it is possible that the educational design of the VLE for mathematics can contribute to the integration of the SAP in the IPGB?

The didactic design was considered as a way to integrate teaching, production and research by the focus groups, which coincided unanimously. The evolution that facilitates the search for a scientific basis and solution for this integration was highlighted.

Operational criteria: unanimity of criteria, given by the consensus existing in the four focus groups and absence of unfavourable or discrepant criteria.

3.2 What would you add or exclude to the educational design of the VLE for the subject of Mathematics to increase its contribution to the integration of the PEA in the IPGB?

In the conception of the didactic design for the desired purpose, three (3) of the focus groups showed consensus by not presenting any limitations; therefore, the majority of the members of focus group 3 placed the need to convert the most graphically enlightening didactic strategy into a scheme that uses a language for representation that is closest to engineers and that will aid their understanding and implementation in the discipline. In the same way, it was decided to go deeper into the description and explanation of the phases of the educational design of the VLE due to its relevance for the future generalization for the whole educational institution.

Operational criteria: most criteria, as three of the four focus groups had consensus from their members.

Agreement 6: to make the didactic strategy more graphically enlightening in a scheme that uses a representation language closer to the engineers and that aids their understanding and didactic implementation in the discipline.

Contract 7: to deepen the description and explanation of the phases of the AVA project due to their relevance for the future generalization in the IPGB.

Theme 4: Coherence between the didactic design of the VLE for the subject of Mathematics and the actual conditions in the IPGB, to contribute to the integration of the PEA.

4.1 Do you consider that the didactic design of the VLE for the subject of Mathematics contains the essential aspects to be taken into account in the IPGB to contribute to the integration of the ASAP?

There was consensus across all groups about the novelty and current nature of linking projects, problems and cases; as well as the integration of technologies used in the two processes. However, four members of Group 2 expressed concern about the principle of didactic design related to technology integration, citing several projects in the IPGB. These criteria were again given a thorough explanation because technology integration was precise and the reason why there was a requirement for project selection as an educational context, associated with project information, was pointed out, which, if not met, would be unique to that project.

Operational criteria: unanimity of criteria, given by the consensus existing in the four focus groups and absence of a majority with unfavourable or discrepant criteria.

Agreement 8: to deepen the explanation of the preparation of the project in the design phases of the VLE to become a formative context of the Mathematics discipline.

4.2 What are the elements for your consideration that may limit the widespread implementation of the VLE didactic design in the IPGB?

More than half of the members of focus group 4 considered susceptible and that it could limit the generalized implementation in the IPGB, understood by specialists or teachers, that the productive issue of the projects would always serve for teaching, whenever the scope of the project does not match or exceeds the teaching purposes. In view of this criterion, it was decided to review the requirements for screening projects related to this subject.

Operational criteria: most criteria, as three of the four focus groups had consensus from their members.

Contract 9: to review the requirements for the screening of the industrial design, contained in its preparation for use as a training context, in the educational design phases of the VLE.

Appendix 11: Questionnaires for the ladov technique.

Appendix 11.1: Student questionnaire.

Dear student, we would appreciate your open and honest collaboration in improving the design and practice of mathematics through your responses to this questionnaire. Please read each question carefully before answering. The questionnaire is anonymous. Thank you very much.

1. Do you like the IPGB? Yes __. No____. I don't know___.
2. What are the three (3) subjects you most enjoy in your career?

a) b) c).

3. Are you satisfied with the subject's AVA? Yes. No. I don't know.
4. What do you like most about the way the subject is taught today?
5. What do you like most about the way the subject is taught today?
6. Do you have any concerns about the matter? Yes. No.

(a) If yes, list your concerns:

7. What are the three (3) subjects you most enjoy in your career?

a) b) c).

8. If you can choose the way you communicate, work collectively and relate to the subject and the software development project, which, for your preference, should be used in the subject, would you choose the VLE?

Yes. No. I don't know.

9. How is the software development project in which you participate? Mark with an X the characteristics that characterize it:

Interesting   Flexible _ Motivating Facilitating_ Organised   .

Bored__ demanding___ demotivating__ challenging__ disorganised__.

10. Do you like the communication spaces and services, the tools for collective work and for relating to the resources that the VLE currently has in the Mathematics course?

a) I like it very much.

b) I like them more than I dislike them.

c) I don't care.

d) I hate them more than I like them.

e) I don't like anything.

f) I can't say.

Annex 12 : Overview of the themes / Workload 11ª Class.
Overview of Subjects / Workload 11ª Class

Unidade de Ensino / Objetivos	Aprendizagem 1 - Conteúdos	Sucessões Sugestões Metodológicas
- Definir sucessão. - Conhecer diferentes formas de definir uma sucessão. - Calcular termos de uma sucessão. - Investigar se um determinado número é ou não termo de uma sucessão. - Estudar a monotonia de uma sucessão. - Calcular, quando existem, os conjuntos de minorantes e majorantes de um conjunto. - Verificar se uma sucessão é limitada. - Definir progressões aritméticas e geométricas. - Calcular a razão de uma progressão. - Escrever o termo geral de uma progressão. - Calcular a soma de n termos consecutivos de uma progressão. - Operar no conjunto $[-\infty, +\infty]$. - Provar que uma sucessão é um infinitésimo e o seu inverso é um infinitamente grande. - Definir limite de uma sucessão. - Levantar indeterminações do tipo 1^∞.	I. Introdução ao conceito de sucessão. 1.1 A sucessão como função de variável natural. 1.2 Sucessões monótonas. 1.3 Sucessões limitadas. 1.4 Progressões aritméticas e geométricas. 2. Limites de sucessões. 2.1 Referência à recta acabada. 2.2 Limites infinitamente grandes e infinitésimos. 2.3 Limite de uma sucessão (convergência); determinação de limites. 2.4 Estudo intuitivo da sucessão $u_n = \left(1 + \frac{1}{n}\right)^n$ como primeira referência de um número e.	- O conceito de sucessão deve ser compreendido como função de domínio IN. - A definição de uma função por recorrência deve ser apresentada em casos simples. - Os alunos devem representar graficamente sucessões (conjunto de pontos isolados de abcissa natural). - Na soma dos m primeiros consecutivos de uma progressão poder-se-á utilizar o símbolo de somatório: $$S_n = \sum_{n=1}^{m} a_n$$ - O aluno deve conhecer as operações básicas [$-\infty, +\infty$]. Sendo a $\in \mathbb{R}^+$ $\pm a + \infty = +\infty$ $\pm a - \infty = -\infty$ $+\infty \cdot a = +\infty$ $+\infty \cdot (-a) = -\infty$ $-\infty \cdot a = -\infty$ $-\infty \cdot (-a) = +\infty$ $\frac{a}{\pm\infty} = 0$ - A definição formal para sucessão convergente para a $\left(\lim u_n = a\right)$ $\forall \delta \in \mathbb{R}^+, \exists p \in \mathbb{N}: n > p \Rightarrow_n |u_n - a| < \delta$ Só deve ser dada como informação. - O conceito de limite de uma sucessão deve ser tratado inicialmente de forma intuitiva. O professor deve pedir que o aluno calcule uma série de termos da sucessão e que conjunture a existência de um valor real para o qual, os termos, se aproximam cada vez mais até que essa diferença seja praticamente nula. - O professor deve esclarecer bem a diferença entre o conceito de sucessão limitada e o conceito de sucessão com limite. - Os limites a pedir não devem exceder o grau de dificuldade de: $\lim\left(1 + \frac{5}{n+2}\right)^{2n}$ Ou $\lim\left(\frac{n+3}{n+5}\right)^{n+2}$

Unidade de Ensino / Aprendizagem 2 - INTRODUÇÃO AO CÁLCULO DIFERENCIAL I		
Objetivos	Conteúdos	Sugestões Metodológicas
- Calcular o domínio de uma função. - Definir assímptotas (vertical e não vertical). - Operar com funções racionais. - Definir função injectiva. - Definir função inversa. - Caracterizar função inversa. - Definir função irracional. - Determinar o domínio e contradomínio de uma função irracional. - Resolver equações irracionais. - Determinar o limite de uma função num ponto. - Levantar indeterminações. - Calcular a taxa de variação média de uma função no intervalo $[a, b]$. - Calcular, usando a definição, a derivada de uma função num ponto. - Calcular as derivadas laterais num ponto da função. - Aplicar regras de derivação da soma, do produto, do quociente e da potência. - Calcular a 2ª derivada de uma função. - Estudar a variação do sinal de uma expressão da derivada de uma função. - Determinar as coordenadas dos pontos do domínio da função em que a 1ª derivada muda de sinal; extremos relativos.	1. Estudo de propriedades das funções racionais, com especial realce para as assímptotas verticais e não verticais. 2. Operações com funções: soma, diferença, produto, quociente e composição. 3. Inversão de uma função. 4. Funções irracionais (índices 2 e 3). 5. Introdução ao conceito de limite segundo Heine. 6. Limite da soma, produto, quociente e potência de funções. Indeterminações. 7. Taxa de variação média; Introdução ao conceito de derivada. 8. Definição geométrica de derivada de uma função num ponto. 9. Derivadas laterais. 10. Regras de derivação. 11. Sinal da primeira derivada e monotonia de uma função; Sinal da 2ª derivada e sentido das concavidades. 12. Extremos relativos de uma função.	- As assímptotas verticais devem ser exploradas a partir do domínio da função racional do tipo $f(x) = \frac{ax+b}{cx+d}$ - Para o cálculo da assímptota horizontal deverá ser usado o algoritmo da divisão e o aluno será levado a concluir que a assímptota tem por equação $y = \frac{a}{c}$ (quociente da divisão). A assímptota não vertical de funções do tipo $g(x) = \frac{ax^2+bx+c}{dx+e}$ também pode ser determinada usando o algoritmo da divisão inteira de polinómios. - O gráfico da função inversa de $f(x)$ deve ser construído pela simetria em relação à bissectriz dos quadrantes ímpares ($y = x$). - As funções irracionais não devem exceder o grau de dificuldade de: $f(x) = 1 + \sqrt{x^2 - 4}$; $f(x) = 2 - \sqrt{x^2 + 9}$; $f(x) = \sqrt[3]{x+2}$ -Usando uma função do tipo $y = e(t)$ poder-se-á pedir a velocidade média num intervalo $[a, b]$. - A derivada de uma função num ponto é o declive da recta tangente ao gráfico da função nesse ponto. $f'(x) = \lim_{x \to x_0} \frac{f(x) - f(x_0)}{x - x_0}$ Poder-se-á falar também na recta normal. - A regra da derivada de $y = \sqrt[n]{f(x)}$ deverá ser deduzida a partir da derivada da potência dado que $y = (f(x))^{\frac{1}{n}}$ - A relação do sinal da 2ª derivada com o sentido das concavidades deve ser dada intuitivamente pela observação de gráficos, podendo-se falar em ponto de inflexão. - O professor deverá dar a informação de que à abcissa do ponto em que a função atinge o máximo se dá o nome de maximizante e à abcissa do mínimo se dá o nome de minimizante.

Unidade de Ensino / Aprendizagem 3 – NO ESPAÇO GEOMETRIA		
Objetivos	Conteúdos	Sugestões Metodológicas
- Marcar um ponto $P(x_\circ, y_\circ, z_\circ)$ num sistema cartesiano, ortogonal e monométrico. - Calcular as coordenadas de um vector, usando a igualdade, $\overline{AB} = B - A$ - Calcular as coordenadas do ponto médio de $[AB]$. - Determinar a distância entre dois pontos. - Calcular o produto escalar de dois vectores. - Escrever equações da recta. - Escrever equações do plano. - Escrever uma equação cartesiana do plano mediador de $[AB]$. - Definir analiticamente superfície esférica e esfera. - Resolver sistemas de três equações com três incógnitas pelos métodos de adição ordenada, substituição, misto e triangulação de Gauss. - Estudar a posição relativa de rectas e planos.	1. O método cartesiano para estudar a geometria no espaço. 2. Vectores livres em $\mathbb{R}^3$. 3. Produto escalar de dois vectores. 4. Equações vectoriais, paramétricas e cartesianas da recta. 5. Equação cartesiana do plano. 6. Equação da superfície esférica; condição que define a esfera. 7. Intersecção de planos; sistemas de 3 equações com três incógnitas. 8. Paralelismo e perpendicularidade de rectas e planos no espaço.	- É importante o aluno aperceber-se das analogias e diferenças entre os conceitos estudados em $\mathbb{R}^2$ e os que serão agora estudados em $\mathbb{R}^3$. - O aluno deve concluir que se M é o ponto médio de $[AB]$, então, $\overline{AM} = \overline{MB}$. - Equações do plano devem ser encontradas usando um vector perpendicular ao mesmo. - A calculadora permitirá rapidamente verificar se os resultados obtidos na resolução do sistema estão correctos.

I want morebooks!

Buy your books fast and straightforward online - at one of world's fastest growing online book stores! Environmentally sound due to Print-on-Demand technologies.

Buy your books online at
www.morebooks.shop

Kaufen Sie Ihre Bücher schnell und unkompliziert online – auf einer der am schnellsten wachsenden Buchhandelsplattformen weltweit! Dank Print-On-Demand umwelt- und ressourcenschonend produzi ert.

Bücher schneller online kaufen
www.morebooks.shop

info@omniscriptum.com
www.omniscriptum.com

Printed by Books on Demand GmbH, Norderstedt / Germany